Electronic
Service
Instruments

Clyde N. Herrick

San Jose City College

San Jose, California

Prentice-Hall, Inc., Englewood Cliffs, New Jersey

Library of Congress Cataloging in Publication Data

HERRICK, CLYDE N.
  Electronic service instruments
  1. Electronic instruments. 2. Electronic measurements. I. Title.
  TK7878.4.H46   621.381'028   73-22123
  ISBN 0-13-251868-6

PRENTICE-HALL SERIES IN ELECTRONIC TECHNOLOGY

Dr. Irving L. Kosow, editor
Charles M. Thomson and Joseph J. Gershon,
consulting editors

© 1974
by PRENTICE-HALL, INC.
Englewood Cliffs, New Jersey

10 9 8 7 6 5 4 3 2 1

Printed in the United States of America

PRENTICE-HALL INTERNATIONAL, INC., *London*
PRENTICE-HALL OF AUSTRALIA, PTY., LTD., *Sydney*
PRENTICE-HALL OF CANADA, LTD., *Toronto*
PRENTICE-HALL OF INDIA PRIVATE LIMITED, *New Delhi*
PRENTICE-HALL OF JAPAN, INC., *Tokyo*

Contents

Preface

More test equipment is used in the home electronics entertainment area than in any other branch of the electronics industry. Servicing of AM, FM, and CB radio receivers, black-and-white and color television receivers, high-fidelity record and tape players, electronic organs, guitar amplifiers, and wireless intercommunication units requires a wide range of both general and specialized electronic test equipment. In turn, the need for a definitive textbook with state-of-the-art coverage of this broad spectrum of electronic instruments is apparent. Instrument design has undergone considerable evolution in recent years, notably with the obsolescence of the electron tube and entry of various solid-state devices.

The volt-ohm-milliammeter still occupies a prominent place in the service industry. However, the vacuum-tube voltmeter has been largely replaced by the transistor volt-ohmmeter. Ohmmeter design has also evolved, and hi-lo ohmmeters are now in wide use. This type of ohmmeter is more informative in analysis of solid-state circuitry than the conventional type of ohmmeter. AM and FM signal generators are utilized, as in the past, for alignment and signal substitution in radio receivers. Nearly all generators of recent manufacture are solid-state designs. Sweep-frequency and marker generators are employed for alignment and circuit analysis in television receivers. There has been a marked trend toward functional design, featuring pushbutton controls.

Oscilloscopes are used in visual-alignment procedures and waveform analysis in both black-and white and color television trouble-shooting. Deisgn sophistication has been much in evidence, with triggered and calibrated time bases supplanting the older free-running time bases. There have also been some inroads by operating features such as dual-trace screen displays and vectorscope facilities. Vertical and horizontal amplifier design has been improved in some oscilloscopes to the extent that practical vectorscope checks can be made in low-level chroma circuits. All present-day service oscilloscopes have a bandwidth of at least 5 MHz.

Audio oscillators and stereo-multiplex generators are basic instruments at the high-fidelity bench. Modern audio generators have a much purer waveform output than their predecessors. The better class of high-fidelity shops also utilize harmonic distortion meters which may be supplemented by intermodulation analyzers. These instruments permit an accurate quantitative evaluation of high-fidelity performance. All of the modern hi-fi service instruments are solid-state in design.

There has also been a trend toward highly specialized television test equipment, such as various television analyzers. These are basically signal-substitution instruments with a very wide range of signal outputs. A typical TV analyzer provides both black-and-white and color test patterns, white-dot and crosshatch patterns, RF, IF, video, and chroma-frequency signals, intercarrier sound signal, audio-test signal, sync, sweep, and AGC signals. Various minor test functions are also provided. TV analyzers are chiefly useful for preliminary trouble localization, and must be supplemented by conventional instruments in most situations to close in on defective components.

This book is intended both for self-study in the home and for classroom use. It is the outcome of extensive teaching experience, not only of the author, but also of his associates at San Jose City College. In a very significant sense, this textbook represents a team effort, and the author would be remiss if he did not acknowledge the important contributions and assistance of the electronics teaching staff. I wish to express my appreciation also to the artist, Robert Mosher. The design and development of this particular text are directed primarily to the needs of the student who is studying at home and does not have an instructor available to guide his progress and answer his questions.

It is assumed that the student has passed previous courses in arithmetic, algebra, geometry, and perhaps trigonometry. Other prerequisites are courses in electricity, basic electronics, and theory of radio, black-and-white, and color television. However, an apt student can study this textbook profitably if he is taking a concurrent course in television theory. It is desirable for the student to have had a previous course in high-fidelity and stereo-multiplexing theory. However, this is not absolutely essential, and most of the important theoretical concepts can be gleaned from careful reading of the pertinent instrument discussions.

CLYDE N. HERRICK

1

Multimeters and Electronic Multimeters

1.1 SURVEY OF MULTIMETERS

We can readily understand that the multimeter is the most useful and the most basic of all service instruments. That is, a voltmeter, current meter, or ohmmeter shows circuit-action facts in their most fundamental form. These facts are presented in quantitative values; in other words, the multimeter is a *measuring* instrument. The units of measurement are the volt, the ampere, and the ohm, any one of which is related to the other two by Ohm's law:

$$\text{Amperes} = \frac{\text{Volts}}{\text{Ohms}} \tag{1.1}$$

$$I = \frac{E}{R} \tag{1.2}$$

It is instructive to consider the operating features of a multimeter, such as illustrated in Fig. 1–1. Its four chief functions are depicted in Fig. 1–2. For example, when the selector switch is set to a DC voltage range, the terminal voltage of a battery can be measured, as shown in Fig. 1–2(a). Or, when the selector switch is set to an AC voltage range, the voltage at an AC wall outlet can be measured, as shown in Fig. 1–2(b). Again, when the selector switch is set to a resistance range, the value of a resistor can be measured, as shown in Fig. 1–2(c). Finally, when the selector switch is set to a DC current range, the instrument can be connected in series with a circuit to measure current values, as shown in Fig. 1–2(d).

1

Fig. 1–1   Typical service-type multimeter (Courtesy of Triplett Elec. Inst. Co.)

Another function of a multimeter is the measurement of AC voltage in the presence of DC voltage. To understand this function, it is helpful to note the AC and DC relationships depicted in Fig. 1–3. We observe that an AC voltage might occur alone. On the other hand, the AC voltage might be mixed with a DC voltage. If the AC voltage does not cross the zero axis, we refer to the mixture as *pulsating DC*. However, if the AC voltage crosses the zero axis, we refer to the mixture as *AC with a DC component*. Whenever we proceed to measure AC voltage in the presence of DC voltage, it is necessary to use the *output* function of the multimeter, as shown in Fig. 1–4. To anticipate subsequent discussion, the output function rejects the DC that is present and passes only the AC.

Conversely, DC voltage can be measured in the presence of AC voltage when a multimeter is set to its DC-voltage function; the meter will not respond to AC voltage on this function. A multimeter that is operating on its DC-voltage function and is measuring power-supply voltage indicates only the DC voltage component for example. The AC voltage component is rejected.

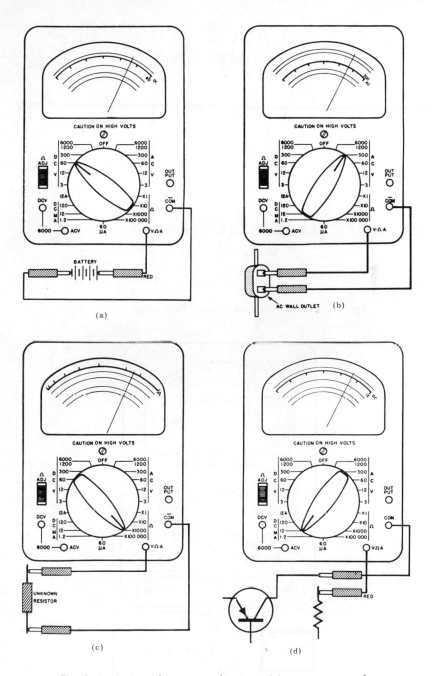

Fig. 1—2   Basic multimeter applications: (a) measurement of
DC voltage; (b) measurement of AC voltage; (c)
measurement of resistance values; (d) measurement
of DC current values (Courtesy of Triplett Elec.
Inst. Co.)

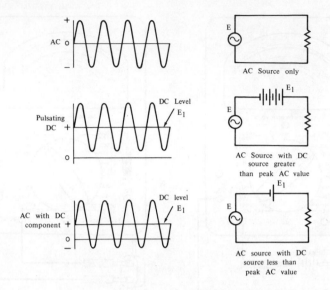

Fig. 1–3   Combinations of AC and DC voltage

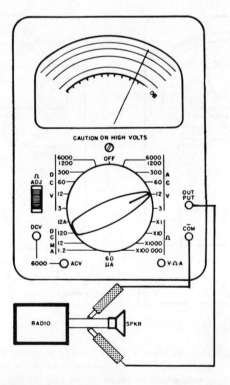

Fig. 1–4   Use of output function to measure AC voltage in the presence of DC (Courtesy of Triplett Elec. Inst. Co.)

1.2   MULTIMETER SCALES

Most multimeters used in electronic service shops are analog types. This means that electrical values are indicated by a pointer on a scale. In practice, a multimeter is provided with several scales, as Fig. 1–5 shows. More than one scale is needed because successive voltage ranges are not decimal multiples of one another. Note in Fig. 1–5 that three DC voltage scales are provided, with full-scale values of 10, 50, and 250, respectively. A separate ohm scale is necessary because its zero point is at the opposite end, compared with a volt scale. We also observe that the ohm scale is highly nonlinear, whereas a DC volt scale is linear. Therefore, resistance values cannot be indicated on a volt scale, nor vice versa. Two AC volt scales are provided, because the lower AC ranges provide somewhat nonlinear indication. A separate decibel scale is necessary because of its considerable nonlinearity and its individual calibration.

It is instructive to observe the range-switch settings that correspond to the various scales in the example of Fig. 1–5. Note in Fig. 1–6(a) that six DC voltage ranges are provided: 2.5, 10, 50, 250, 1000, and 5000 volts full scale. When operating on the 2.5-volt range, we read the 250-volt scale and shift the decimal point two places to the left. When operating on the 10-volt range, the 10-volt scale is read directly. Similarly, when operating on the 50-volt range, the 50-volt scale is read directly. Again, when operating on the 250-volt range, the 250-volt scale is read directly. However, when operating on the 1000-volt range, we read the 10-volt scale and shift the decimal point two places to the right. In the same manner, when operating on the 5000-volt range, we read the 50-volt scale and shift the decimal point two places to the right. Note also that the same range-switch setting is used on both the 1000-volt and the 5000-volt ranges—we move the test lead from "+" to the "DC 5000 V" pin jack to change ranges.

Another essential operating feature concerns the setting of the AC-DC function switch in Fig. 1–6. In this example the switch is set to "+DC." In turn, the pointer will deflect upscale when the "Common" or "−" lead is connected to the negative terminal of a potential source, and the "+" lead is connected to the positive terminal of the potential source. If the test leads are reversed, the pointer will deflect downscale. However, if the function switch is then set to "−DC," the pointer will again deflect upscale. This is a very practical consideration when measuring voltage values in electronic circuits, as shown in Fig. 1–7. That is, a PNP transistor operates with reversed terminal polarities, compared with an NPN transistor. Therefore, it is a considerable operating convenience to employ a multimeter with a polarity-reversing switch. This facility avoids the necessity for frequent reversal of test leads in practical servicing procedures.

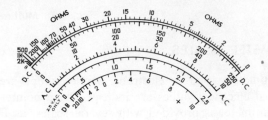

Fig. 1–5   Scale plate for a typical multimeter (Courtesy of Simpson Electric Co.)

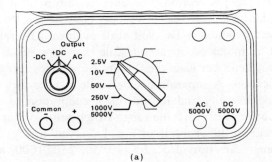

(a)

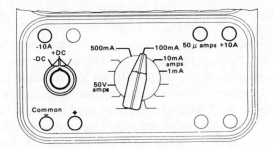

(b)

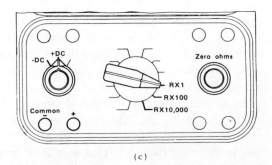

(c)

Fig. 1–6   DC voltage, current, and resistance ranges for Fig. 1–5: (a) voltage ranges; (b) current ranges; (c) resistance ranges

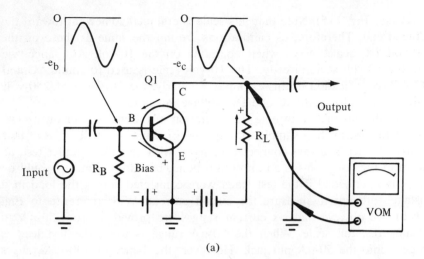

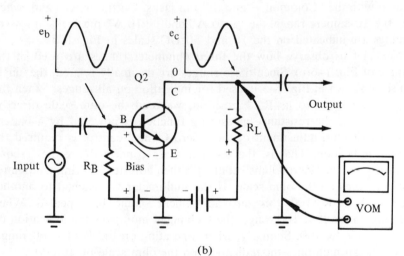

Fig. 1–7   Example of reverse polarities in similar circuits: (a)
PNP transistor operates with negative collector voltage; (b) NPN transistor operates with positive
collector voltage

Next, let us consider the scales that are employed on various AC voltage
ranges in the example of Fig. 1–5. With reference to Fig. 1–6(a), we set
the function switch to its "AC" position, and use the same range-switch
settings as for DC voltage measurements. However, a different 2.5-volt
scale is required to obtain maximum accuracy of indication. Thus, when
operating on the 2.5-volt AC range, we use the scale marked "2.5 V AC

Only" (see Fig. 1–5). Note that this scale is cramped somewhat toward the lefthand end. Therefore, its calibrations are not the same as those on the 250-volt DC scale. Next, when operating on the 10-volt AC range, we read the 10-volt scale directly. The 10-volt scale is used for both AC and DC voltage indication in this example. Similarly, the 50-volt and 250-volt scales are used for both AC and DC voltage indication.

It is also instructive to note the operation of the multimeter on its DC current ranges, in the example of Fig. 1–6(b). First, we will observe that a current meter is always connected in series with the circuit under test, as shown in Fig. 1–8. This is an important point, because the meter would be seriously overloaded if the test leads were connected across the load in a mistaken attempt to measure the circuit current. With reference to Fig. 1–6(b), we observe that six current ranges are provided, from 50-μA to 10 amperes, full scale. When the 50-μA range is used, the test lead is plugged into the 50-μA pin jack. However, the 1-mA to 500-mA ranges are used with the "Common − and +" pin jacks. On the other hand, when using the 10-ampere range, the "−10 A" and "+10 A" pin jacks are used. Readings are indicated on the 10 and 50 DC scales in Fig. 1–5.

Next, let us observe how the three ohmmeter ranges are used in the example of Fig. 1–6(c). Since these ranges are decimally related, the single ohm scale shown in Fig. 1–5 is used for indication on all ranges. When the range switch is set to its R×1 position, we read the ohm scale directly. Note that since the resistance-measuring function is powered by a battery in the multimeter, adjustment of the "Zero Ohms" control is required for accurate indication. That is, the test leads are connected together (short-circuited), and the "Zero Ohms" control is then adjusted to bring the pointer exactly to zero on the ohm scale. If the multimeter is switched to another range, such as R×100, this zero adjustment should be repeated. When operating on the R × 100 range, the technician multiplies the indication on the ohm scale by 100. Similarly, when operating on the R×10,000 range, the technician multiplies the indication on the ohm scale by 10,000.

We will find that some resistors, such as composition and wirewound units, are linear and bilateral. In other words, the resistance reading on the multimeter is the same on each range. Also, the resistance reading remains the same when the test leads are reversed. On the other hand, other resistive components, such as semiconductor diodes and transistors, are nonlinear and unilateral. For example, when the resistance of a semiconductor diode is measured, as shown in Fig. 1–9, we might observe an indication of 3000 ohms in one direction of current flow, and an indication of 4 megohms in the other direction of current flow. This is an example of practical unilateral conductivity. The ratio of forward resistance (3000

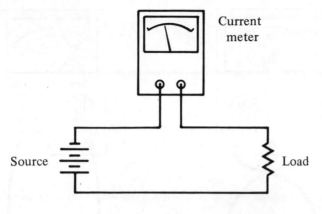

(a)

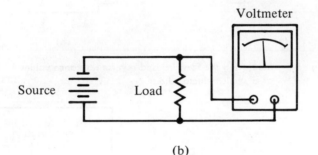

(b)

Fig. 1–8   Test connections for current measurement and for voltage measurements: (a) series connection for current measurement; (b) parallel connection for voltage measurement

ohms) to back resistance (4 megohms) is called the front-to-back ratio (1 to 1333) of the diode under test. Note that if the front-to-back ratio is next measured on another range of the ohmmeter, another ratio will be obtained because a semiconductor diode is a nonlinear resistor. This is just another way of saying that the resistance value of the diode changes when the applied voltage is changed.

An ohmmeter test is often used in preliminary checking of semiconductor devices. In general, a single-junction diode normally shows a high front-to-back ratio. Ordinary transistors have three terminals, as in the example of Fig. 1–10. Note that this transistor design utilizes the

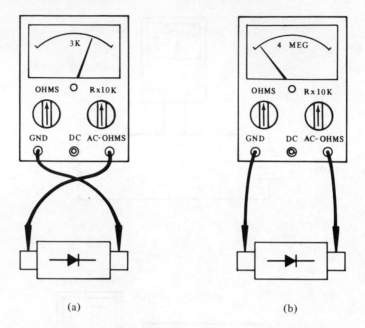

Fig. 1–9   Measuring the forward and reverse resistance values
of a semiconductor diode

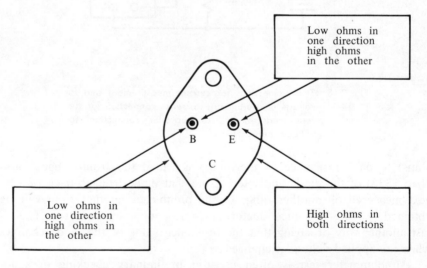

Fig. 1–10   Ohmmeter test of a power transistor

case as the collector terminal. If the transistor is in normal operating condition, an ohmmeter test will show the resistance ratios noted in the diagram.
A defective transistor often shows an open circuit or a short circuit instead

of a normal front-to-back ratio in the base-to-collector or the base-to-emitter test. A poor front-to-back ratio in these tests is also an indication of a defective transistor.

1.3   MULTIMETER PROBES

Probes are used to extend the ranges and functions of a multimeter. One of the most widely used types is the high-voltage DC probe used to measure the picture-tube accelerating voltage in TV receivers. Fig. 1–11 illustrates a typical high-voltage probe. The probe is designed to provide a specified full-scale voltage indication when used on a certain DC voltage range of a multimeter that has a specified sensitivity. For example, a high-voltage probe might be rated for a full-scale indication of 25,000 volts when used on the 1000-volt DC range of the multimeter shown in Fig. 1–6(a). In such a case, we would read the indication on the 250-volt scale (Fig. 1–5), and multiply the reading by 100. Next, let us consider why the multimeter must be operated on a specific range, and why the instrument must have a specified sensitivity.

In the example under discussion, the multimeter has the circuitry shown in Fig. 1–12 on its DC-voltage function. This configuration operates at a sensitivity of 20,000 ohms-per-volt. The input resistance of the multimeter on any DC range setting is equal to 20,000 times the full-scale value of that range. For example, if we operate on the 2.5-volt DC range, the instrument has an input resistance of 50,000 ohms. Or, if we operate on the 1000-volt DC range, the input resistance is 20 megohms. Since a high-voltage probe is basically a voltage divider (external multiplier resistor), it follows that the probe can be designed for proper scale indication only when used on a specified DC voltage range of the multimeter.

Fig. 1–11   High-voltage DC probe (Courtesy of B & K Mfg.
Co., Division of Dynascan Corp.)

Note that each of the resistors in the configuration of Fig. 1–12 is called a multiplier resistor. This term means that each resistor multiplies the full-scale voltage that is required for operation of the meter movement or mechanism. Since the meter movement in this example has an internal resistance of 2000 ohms and requires 50 microamperes for full-scale deflection, it follows that the voltage drop across the meter movement at full-scale deflection is 100 millivolts, in accordance with Ohm's law. Next, when the range switch is set to its 2.5-volt position, a 48K resistor is connected in series with the meter movement. The total input resistance is now 50,000 ohms. Thus, when 2.5 volts are applied, the current flow is 50 microamperes in accordance with Ohm's law. As before, the voltage drop across the meter movement is 100 millivolts.

When the high-voltage probe is connected to the multimeter, it is evident that the probe resistance must be suitable to provide a current flow of 50 microamperes when 25,000 volts are applied to the probe. To find the required value of probe resistance, we observe in Fig. 1–12 that the multimeter has an input resistance of 20 megohms on its 1000-volt DC range. Therefore, the probe must contain a multiplier resistance of 480 megohms. The total input resistance of the probe and multimeter will be 500 megohms. In turn, the current flow at 25,000 volts is 50 microamperes.

$$I \; = \; \frac{25,000}{500 \times 10^6} = 50 + 10^{-6} \text{ ampere} \tag{1.1}$$

The student may show that if a multimeter utilizes a meter movement that provides full-scale deflection at 10 microamperes, the instrument sensitivity will be 100,000 ohms-per-volt. (*Hint*: It follows from previous discussion that the sensitivity of a multimeter is equal to the reciprocal of the full-scale current value on any DC-voltage range, in accordance with Ohm's law.)

$$\text{Ohms-per-volt} \; = \; \frac{1}{I_{\text{F.S.}}} \tag{1.2}$$

where $I_{\text{F.S.}}$ is the full-scale current value of the meter movement.

Another useful multimeter probe is the RF diode probe, also called a signal-tracing probe. A typical configuration is shown in Fig. 1–13. It is generally operated on the first DC-voltage range of a multimeter. This type of probe extends the AC-voltage function of a multimeter, inasmuch as the probe can be used at frequencies up to 200 MHz. On the other hand, when a multimeter is used on its AC-voltage function with ordinary test leads,

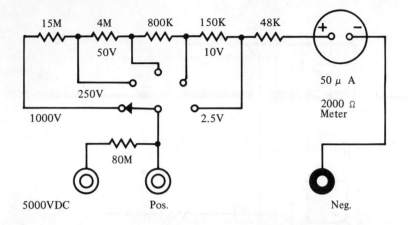

Fig. 1—12 Configuration utilized on the DC voltage function of a multimeter

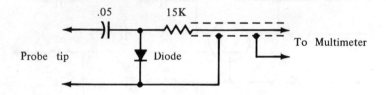

Fig. 1—13 Signal-tracing probe for a multimeter

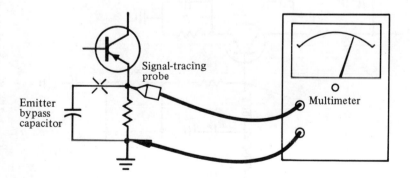

Fig. 1—14 Check of emitter bypass capacitor

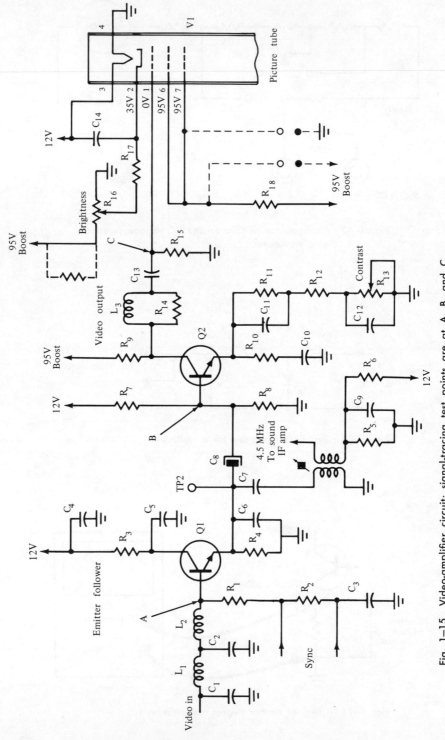

Fig. 1–15 Video-amplifier circuit; signal-tracing test points are at A, B, and C.

14

the response drops off rapidly above the audio-frequency range. Note that a signal-tracing probe is not intended to provide accurate RF-voltage measurements, but is used primarily to check for presence or absence, and for comparative levels of high-frequency voltages. It is instructive to consider several types of service tests made with a signal-tracing probe and multimeter.

With reference to Fig. 1–14, a signal-tracing probe can be used to check for open emitter bypass capacitors. An open capacitor results in low gain. When an emitter bypass capacitor opens, the signal voltage at the emitter terminal rises greatly. In a typical amplifier stage, the emitter signal voltage increases approximately 10 times if the emitter bypass capacitor is open. Therefore, an open capacitor can be quickly and easily located with a signal-tracing probe and multimeter. Note that tests can be made in comparatively low-level stages if a signal-tracing probe is used with a high-sensitivity multimeter, such as a 100,000 ohms-per-volt instrument. The probe has least utility with a low-sensitivity multimeter, such as a 1000 ohms-per-volt instrument.

An open coupling capacitor, such as in the video amplifier of a television receiver, stops the signal and causes a "no picture" symptom. For example, if capacitor C8 or C13 in Fig. 1–15 opens, the signal will be stopped. To check for an open coupling capacitor, the signal-tracing probe is applied first at A. In case a signal indication is found here, a test is made next at B. No signal indication at B means that C8 is probably open. To confirm this possibility, we check for signal at TP2. If a signal indication is found here, it is logical to conclude that C8 is open. However, if there is no signal at TP2, Q1 is likely to be found defective. On the other hand, in case a signal indication is found at B, we proceed to check for signal at C. No signal indication at C means that C13 is probably open. To confirm this possibility, we check for signal at the collector of Q2. If a signal indication is found here, it is logical to conclude that C13 is open. However, if there is no signal at the collector of Q2, the transistor is likely to be found defective.

A signal-tracing probe can also be used with a multimeter to check the operation of Citizen Band (CB) or amateur-radio transmitters. Thereby, adjustments of power-output and matching circuits and antenna checks are facilitated. The procedure is: A four-foot lead is connected to the tip of the signal-tracing probe, to serve as a pick-up antenna. The multimeter is operated on its lowest DC-voltage or current range. Then the pick-up lead is positioned to give approximately half-scale indication on the meter. Thereafter, the position of the pick-up lead is not changed. Transmitter and antenna adjustments are made to obtain maximum signal indication. In this application, the signal-tracing probe and multimeter are being used as a relative field-strength meter.

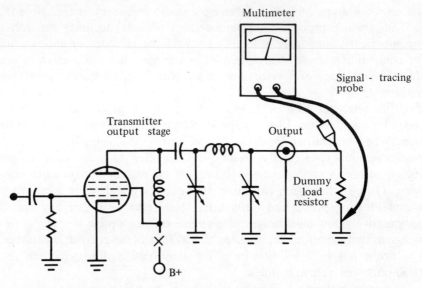

Fig. 1–16   Test setup for checking transmitter adjustments

A signal-tracing probe and multimeter are also very useful for checking CB or amateur-radio transmitter neutralizing adjustments. If a conventional pi-network output stage is used, the test setup shown in Fig. 1–16 is suitable. With the antenna disconnected from the transmitter, a non-inductive load resistor having a value equal to the antenna impedance is connected across the output terminals. The antenna impedance is usually 50 or 72 ohms. Short leads should be used to connect the dummy load resistor to the output terminals. Connect the signal-tracing probe across the dummy load resistor. To test the neutralizing adjustments, the plate and screen-grid power supplies are disconnected from the output stage, but the normal grid drive is applied. Then the neutralizing circuit is adjusted for a minimum meter reading. Trim up the grid and plate-tuned circuits for any increase in the reading that might be obtainable, and then repeat the adjustment of the neutralizing circuit for a minimum reading.

1.4   OUTPUT FUNCTION AND DECIBEL INDICATION

As mentioned previously, the output function of a multimeter is the same as its AC-voltage function, except that any DC component that might accompany that AC is rejected. For example, with reference to Fig. 1–3, the output function is used to convert pulsating DC to an AC waveform. Thereby, accurate measurements of AC voltage are obtained in the presence of DC voltage. A practical example is seen in Fig. 1–7, wherein the AC voltage at the collector of the transistor is accompanied by the DC supply voltage. Therefore, to measure the AC collector voltage with a multi-meter, it is necessary to use the output function of the instrument. Note that most of the AC voltages in an audio amplifier are accompanied by DC supply or bias voltages.

When checking stage gains and signal levels in audio amplifiers, we often read the decibel scale instead of the AC voltage scale of a multimeter.

Power ratio	Voltage ratio	dB - +	Voltage ratio	Power ratio
1.000	1.0000	0	1.000	1.000
.9772	.9886	.1	1.012	1.023
.9550	.9772	.2	1.023	1.047
.9333	.9661	.3	1.035	1.072
.9120	.9550	.4	1.047	1.096
.8913	.9441	.5	1.059	1.122
.8710	.9333	.6	1.072	1.148
.8511	.9226	.7	1.084	1.175
.8318	.9120	.8	1.096	1.202
.8128	.9016	.9	1.109	1.230
.7943	.8913	1.0	1.122	1.259
.6310	.7943	2.0	1.259	1.585
.5012	.7079	3.0	1.413	1.995
.3981	.6310	4.0	1.585	2.512
.3162	.5623	5.0	1.778	3.162
.2512	.5012	6.0	1.995	3.981
.1995	.4467	7.0	2.239	5.012
.1585	.3981	8.0	2.512	6.310
.1259	.3548	9.0	2.818	7.943
.1000	.3162	10.0	3.162	10.00
.07493	.2818	11.0	3.548	12.59
.06310	.2512	12.0	3.981	15.85
.05012	.2293	13.0	4.467	19.95
.03981	.1995	14.0	5.012	25.12
.03162	.1778	15.0	5.623	31.62
.02512	.1585	16.0	6.310	39.81
.01995	.1413	17.00	7.079	50.12
.01585	.1259	18.00	7.943	63.10
.01259	.1122	19.0	8.913	79.43
.01000	.1000	20.0	10.001	100.00
10^{-3}	3.162×10^{-2}	30.0	3.162×10	10^{-3}
10^{-4}	10^{-2}	40.0	3.162×10^{2}	10^{4}
10^{-5}	3.162×10^{-3}	50.0	3.162×10^{2}	10^{5}
10^{-6}	10^{-3}	60.0	10^{3}	10^{6}
10^{-7}	3.162×10^{-4}	70.0	3.162×10^{3}	10^{7}
10^{-8}	10^{-4}	80.0	10^{4}	10^{8}
10^{-9}	3.162×10^{-5}	90.0	3.162×10^{4}	10^{9}
10^{-10}	10^{-5}	100.0	10^{5}	10^{10}

Fig. 1–17   Corresponding power ratios, voltage ratios, and dB values

Refer to Fig. 1–5 for the relation of a decibel (dB) scale to the AC scales. It is preferred to measure audio signals in dB, because dB values are proportional to ear response. On the other hand, AC voltage levels are not proportional to the loudness of an audio signal. We will find that dB values are based on power ratios, and that dB measurements are additive and subtractive. For example, if we measure a loss of 20 dB through a volume control, and measure a gain of 30 dB through the following amplifier stage, the over-all gain is then 10 dB. Now let us consider the correct way to measure dB with a multimeter.

The fact that dB values are power ratios is apparent in the tabulation shown in Fig. 1–17. However, a multimeter cannot measure power values. For this reason, we must measure AC voltage values which correspond to power values. It follows from Ohm's law that an AC voltage ratio corresponds to a power ratio only when the AC voltage measurements are made across the same value of impedance in each case. Moreover, dB readings on a multimeter are correct only when the AC voltage measurements are made across a specified value of impedance. For example, in Fig. 1–5 the dB scale is calibrated with respect to measurements across 600 ohms. In spite of these considerations, a multimeter can be used to make correct dB measurements across loads with various impedances, as follows.

First, we should note that the dB scale on a multimeter is calibrated for direct reading on the first AC-voltage range. Therefore, when we operate the instrument on higher AC-voltage ranges, we must add certain numbers of dB to the reading on the dB scale. In most cases, the numbers of dB that must be added in each voltage range are printed on the meter scale plate. Thus, in the example of Fig. 1–5, if we are operating on the 10-volt AC range, we must add 12 dB to the scale reading. Or, if we are operating on the 50-volt AC range, we must add 26 dB to the scale reading.

Next, we must take into consideration the values of load impedance across which we are measuring dB values. There are three basic situations to be recognized. If load impedances at the input and output of an amplifier are the same as the reference value for the multimeter (600 ohms in this example), the dB gain can be found by subtracting the indicated values at input and output. For example, if we measure 2 dB at the input of the amplifier, and measure 13 dB at the output of the amplifier, the gain is equal to 11 dB. Or, if we measure −1 dB at the input and measure 12 dB at the output, the gain of the amplifier is 13 dB, as before. In other words, when negative dB are involved, we add the measured values algebraically.

However, it seldom happens that amplifier load impedances are equal, or that one of them might be equal to the reference value for the multimeter. Therefore, we evaluate dB scale readings as follows. In case a pair of load impedances are equal in value, but are not the same as the reference value for the multimeter, we have the situation shown in Fig. 1–18(a) where we find the dB gain by simply subtracting the two readings. In other words, the gain in this example is equal to 13 dB. Note carefully that neither the 2-dB reading nor the 15-dB reading is correct of itself—but their difference is nevertheless correct because both measurements were made across the same value of impedance, viz., 75 ohms.

In case the pair of impedances are unequal, we start by subtracting the two readings, but we must then make a final correction from a table. For example, in Fig. 1–18(b), the difference between the two dB readings is 10 dB. In turn, it might appear that there is an over-all loss of 10 dB. However, since the ratio of input and output impedances is 10-to-1, we must check the table in Fig. 1–19. We observe that when the impedance ratio

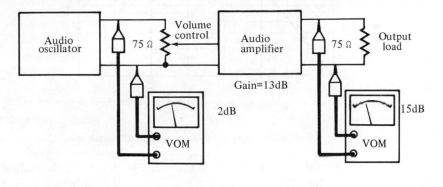

(a)

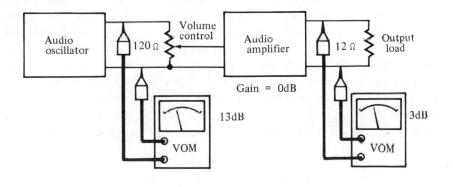

(b)

Fig. 1–18   Decibel measurements in two load-value situations:
(a) input and output impedances equal; (b) input
and output impedances unequal

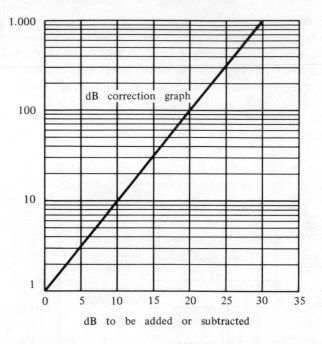

Fig. 1—19 Decibel correction table for unequal impedances

is 10, there are 10 dB to be added or subtracted. In this case, the output impedance is less than the input impedance, so we will subtract 10 dB from the difference that we started with. In other words, we subtract 10 from 10, and conclude that the over-all gain of the amplifier system is zero.

Finally let us consider an amplifier arrangement similar to that in Fig. 1—18(b), but with the input and output impedances reversed. That is, we will stipulate that the input impedance is 12 ohms, the output impedance 120 ohms, with 3 dB measured at the input, and 13 dB measured at the output. In this situation, it might seem that a gain of 10 dB is obtained. On the other hand, when we check the table in Fig. 1—19, we find that there are 10 dB to be added or subtracted. Since the output impedance is greater than the input impedance in this case, we subtract the 10 dB factor. In other words, 10 dB minus 10 dB gives us zero over-all gain, as before.

1.5 CURRENT FUNCTION CIRCUITRY

Previous description of the voltage function circuitry for a multimeter was given. Using the same example, Fig. 1—20 shows the current function configuration. Each resistor is called a meter shunt, and the circuit employs a ring-shunt configuration. The advantage of a ring shunt is that the meter

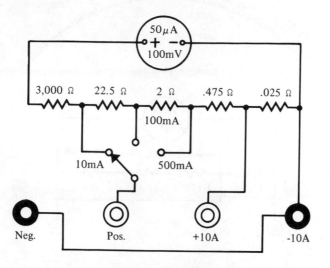

Fig. 1–20   Basic ring-shunt configuration with four current ranges

circuit is never opened while switching from one range to another. In turn, there is less danger of overloading and damaging the meter movement. We observe that each switch position corresponds to a range that provides 50 microamperes full-scale current flow to the meter movement. Since the internal resistance of the movement is 2000 ohms, the voltage drop across the movement is 100 millivolts at full-scale deflection.

To obtain a 50-microampere current range, a separate input terminal may be provided for the switch, so that only the meter movement is in the circuit. Note in Fig. 1–20 that a separate terminal is provided for the 10-ampere range. Because this is a heavy-current range, it is not practical to pass the incoming current through the range switch. That is, the contact resistance of the switch would be excessive and impose undue error. We observe that the input resistance of the current-measuring configuration is comparatively low, and that it varies from one range to another. Thus, the input resistance to 0.5 ohm on the 500-mA range, and 25 ohms on the 10-mA range.

1.6   OHMMETER FUNCTION CIRCUITRY

Figure 1–21 depicts ohmmeter circuitry for R×1, R×100, and R×10,000 resistance ranges. An internal 1.5-volt battery is used on the R×1 range, and the total input resistance is 12 ohms when the lead resistance and the internal resistance of the battery are taken into account. Therefore, the center-scale indication on the R×1 range is 12 ohms.

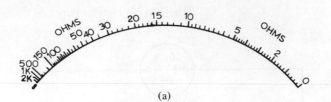

(a)

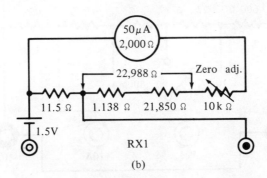

(b)

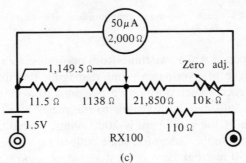

(c)

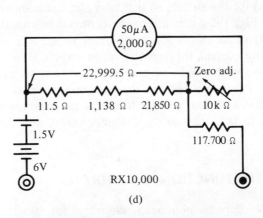

(d)

Fig. 1–21 Typical ohmmeter configurations on various ranges:
(a) scale used on all ranges; (b) meter circuitry on
RX1 range; (c) meter circuitry on RX100 range; (d)
meter circuitry on RX10,000 range

Next, we observe that the 1.5-volt battery is also used on the R×100 range, and that the total input resistance is 1200 ohms. In turn, the center-scale indication on the R×100 range is 1200 ohms. However, since the total input resistance is 120,000 ohms on the R×10,000 range, additional internal battery voltage must be employed to pass 50 microamperes of current through the meter movement when the test leads are short-circuited. Therefore, a 6-volt battery is utilized in series with the 1.5-volt battery on the R×10,000 range.

1.7   AC VOLTAGE FUNCTION CIRCUITRY

A typical configuration for the AC voltage function of a multimeter is shown in Fig. 1–22. We observe that a multiplier circuit is provided in much the same manner as for the DC voltage function. However, diodes CR1 and CR2 are connected in the meter circuit. Diode CR1 provides rectification of AC voltages, so that the meter movement is energized by half-cycles only. Diode CR2 provides improved indication accuracy and also prevents excessive reverse voltage from being applied to CR1 when operating on the higher ranges. That is, CR2 minimizes leakage current through CR1 by providing a low-resistance shunt path during the negative half cycle. Similarly, CR2 prevents the drop of an appreciable potential across CR1 during the negative half cycle. Resistor R is called a swamping resistor. Its shunting action improves the scale linearity to some extent, at the cost of reduced meter sensitivity. Resistor R1 is a calibrating resistor, with a value chosen to provide maximum full-scale accuracy.

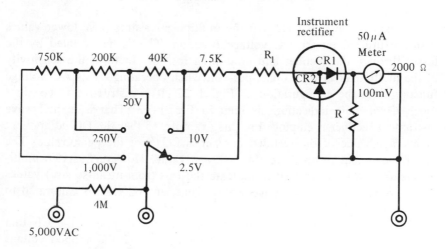

Fig. 1–22   Typical AC voltage configuration for a multimeter

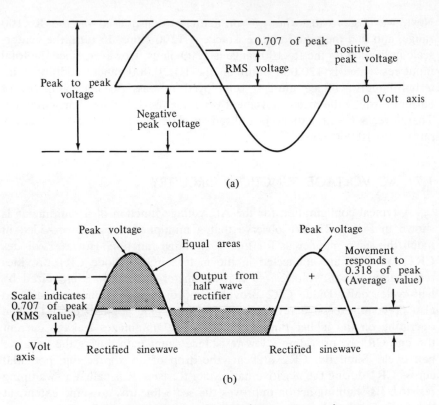

Fig. 1–23 Basis of AC voltage indication in a multimeter:
(a) fundamental levels in a sine wave; (b) fundamental levels in a half-wave rectifier output

We observe in Fig. 1–22 that the multiplier resistors have lower values than are utilized on the DC voltage function. This is necessitated by the losses through the rectifier and swamping resistor. In turn, the sensitivity of the multimeter is 1000 ohms-per-volt when operating on its AC voltage function with the configuration of Fig. 1–22. It is instructive to consider the basis of scale indication, as seen in Fig. 1–23. That is, a half-wave instrument rectifier is employed in this example, so that the DC component or average value of the rectified current is 0.318 of the peak value. The meter movement responds to this average value. However, service multi-meters are always calibrated to indicate the rms (root-mean-square) values of sine waves on the AC voltage scales. Thus, an AC scale is calibrated to indicate rms voltages, which are 0.707 of peak.

A practical consequence of the relations depicted in Fig. 1–23 is that correct AC-voltage values will be indicated only when the applied voltage has a sine waveform. In other words, a multimeter will be quite inaccurate if used to measure the voltage of a square wave, sawtooth wave, or pulse.

The same precaution applies to application of the output function of a multimeter. For example, if dB measurements are made at the input and output of an amplifier that has appreciable distortion, the output measurement will be in error. This is called a form-factor error, inasmuch as a distorted sine wave does not have a true sinusoidal waveform.

1.8   TRANSISTOR VOLTMETERS (TVMs)

A transistor voltmeter is similar in functions to a multimeter. However, due to its internal transistor amplifier, a TVM is more sensitive. The appearance of a typical transistor voltmeter that employs field-effect transistors is shown in Fig. 1–24. A simplified schematic diagram of the DC-voltage function configuration is depicted in Fig. 1–25. The DC voltage ranges are from 1 to 1000 volts, with an input resistance of 15 megohms on all ranges. We observe that the DC amplifier comprises a pair of field effect transistors (FETs) Q1 and Q2 in a balanced bridge circuit. In turn, there

Fig. 1–24   Typical transistor voltmeter (Courtesy of Sencore)

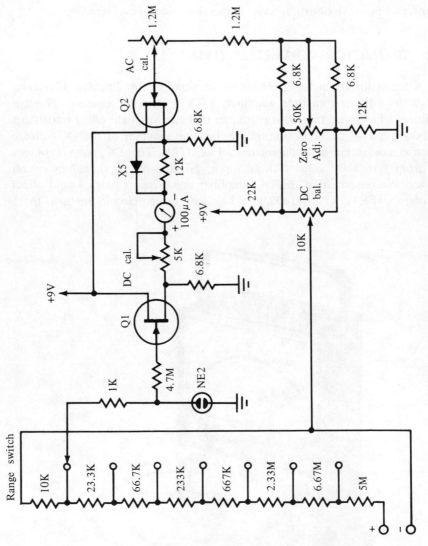

Fig. 1-25 Simplified schematic diagram of TVM DC voltage function circuitry

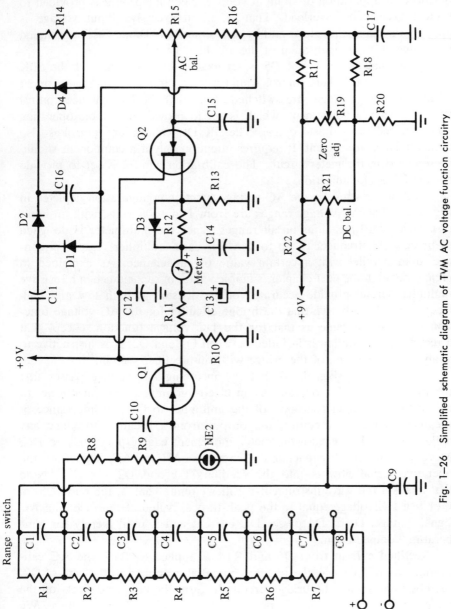

Fig. 1–26 Simplified schematic diagram of TVM AC voltage function circuitry

is no current flow through the meter movement unless the bridge is un-balanced by application of an input voltage. Neon lamp NE2 is provided to protect Q1 against overload. That is, if an excessive input voltage is applied, the neon lamp glows and bypasses the FET. Diode CR5 is included for temperature compensation of the amplifier gain.

Bridge balance in Fig. 1–25 is set exactly by adjustment of the 50K zero-adjust control. This control sometimes requires a slight readjustment when DC voltage ranges are switched. Note that the DC balance control is a maintenance adjustment, which is set to provide a suitable operating point for the zero-adjust control. The 5K DC calibration control is also a maintenance adjustment. It requires attention only if a component should be replaced in the meter circuit. The calibration control is set to provide correct full-scale indication.

Next, let us consider the AC voltage function configuration depicted in Fig. 1–26. The AC voltage ranges are from 1 to 1000 volts, with an input resistance of 10 megohms on all ranges. Since the multimeter is designed to process AC frequencies up to 1 MHz, each multiplier resistor is con-nected in parallel with a certain value of capacitance. As explained in greater detail in the next chapter, suitable values of capacitance in a resistive multiplier circuit provide accurate voltage division at both low and high frequencies. We observe that the bridge circuitry for the AC voltage func-tion is much the same as that for the DC voltage function, except that a rectifier configuration is included between Q1 and Q2. It is instructive to analyze the operation of the bridge with this rectifier section.

We observe in Fig. 1–26 that the incoming AC voltage passes first through Q1. The FET operates as an electronic impedance transformer to match the high input resistance of the multiplier to the low impedance of the meter circuit. Of course, the output from Q1 cannot produce any deflection on M1, because the meter movement can respond only to DC. Moreover, the 10-μF capacitor C13 bypasses M1 for AC. Therefore, the operative signal flow is into the rectifiers D1 and D2 via C11. Note that this is a full-wave instrument-rectifier circuit. That is, the output from C11 is a DC voltage equal to the peak-to-peak value of the incoming AC signal voltage. Diode D4 passes DC current and is used merely for tem-perature compensation.

Rectified current from D4 and R14 is applied to R15, the AC cali-bration control. Thence, the rectified potential is applied to the gate of Q2, thereby unbalancing the bridge circuit. In turn, the pointer deflects on the meter scale by an amount proportional to the value of gate voltage. We recognize in this example that the meter movement responds to the peak-to-peak value of the incoming signal voltage. The AC voltage scales are calibrated in peak-to-peak units, and also in rms units. It is evident that the rms readings will be accurate only when a sine-wave signal is applied

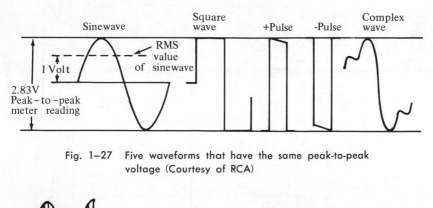

Fig. 1–27   Five waveforms that have the same peak-to-peak voltage (Courtesy of RCA)

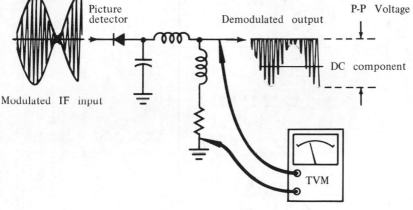

Fig. 1–28   Example of TVM response on AC and DC voltage functions

to the instrument. On the other hand, the peak-to-peak voltage readings will be correct for any waveform. For example, each of the waveforms depicted in Fig. 1–27 will produce the same peak-to-peak reading (2.83 volts p-p), and this is a correct indication. On the other hand, each of the waveforms will produce a different rms reading, and only the sine-wave reading (1 volt rms) will be correct.

Since a series blocking capacitor is provided in the multiplier circuit of Fig. 1–26, the DC component of an applied signal cannot enter the meter circuit. In other words, the TVM operates in a similar manner to an output meter, and only the AC component of the applied signal voltage is passed. It is helpful to observe the application of a TVM in the picture-detector circuit of a TV receiver, as depicted in Fig. 1–28. The output signal from the picture detector has an AC component and a DC component. When the TVM is switched to its AC function, we measure the peak-to-peak voltage of the AC output from the detector. On the other

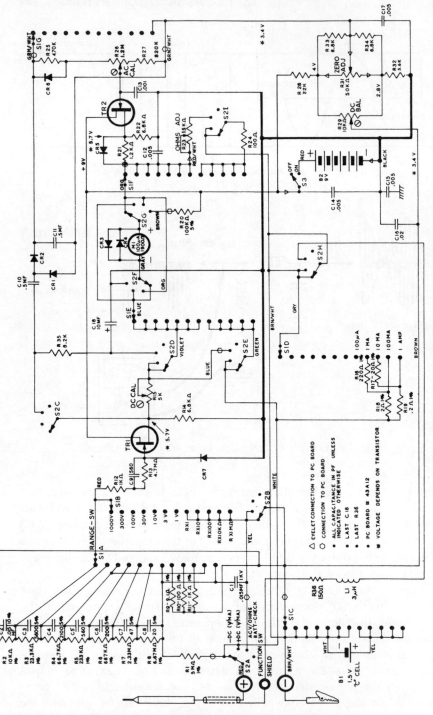

Fig. 1–29 Complete schematic diagram for a TVM (Courtesy of Sencore)

hand, when the TVM is switched to its DC function, we measure the value of the DC component in the output from the detector.

Although the TVM in this example provides DC current ranges, the circuitry that is employed on this function is the same as in a multimeter. In other words, the FET bridge circuit is not used when the instrument is operated on its DC current function. However, the ohmmeter function employed in this example does utilize the bridge circuit. Otherwise, the ohmmeter circuitry is the same as in a multimeter. That is, the output from the internal battery and multiplier are applied to the bridge circuit in the same manner as if DC voltage were being measured. Resistance readings are observed on the ohm scale, which serves the three resistance ranges in the same manner as in a multimeter. A complete circuit diagram for the TVM discussed in this example is given in Fig. 1–29. Note that diodes CR3 and CR4 are included to provide protection for the meter movement in case of accidental overload.

It is important to note the chief advantage of a TVM, compared with a multimeter, as exemplified in Fig. 1–30. We will find that the loading effect and consequent measuring error is much less with the TVM, when testing in high-resistance circuits. When two 500K resistors are connected across a 100-volt source, it follows from Ohm's law that there will be 50 volts

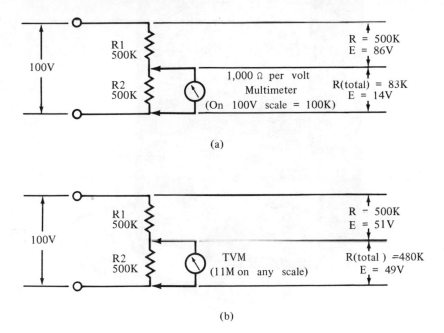

(a)

(b)

Fig. 1–30 Examples of circuit loading: (a) With a 1000 ohms-per-volt meter; (b) With an 11-megohm TVM

drop across each resistor. This is a high-resistance circuit, and we will not obtain the same voltage readings with the multimeter and the TVM. As shown in the diagram, the multimeter in this example reduces R2 to an effective value of 83K, because of circuit loading. In turn, the multimeter reads 14 volts instead of 50 volts. On the other hand, the TVM in this example reduces R2 to an effective value of 480K; thus, circuit loading is minimized. In turn, the TVM reads 49 volts, incurring an error of only 2%.

An important specialized TVM, called the field-effect multimeter, is illustrated in Fig. 1–31. This instrument provides considerable advantage in troubleshooting solid-state circuitry. It has an ohmmeter section with conventional testing facilities, and also has an ohmmeter section with low-

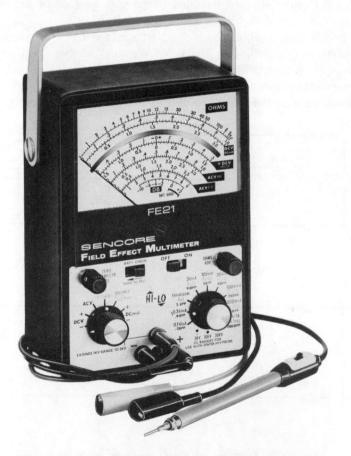

Fig. 1–31   Appearance of the Sencore FE21 Hi-Lo FETVM
(Courtesy of Sencore)

Low voltage of .08 V prevents transistor from conducting and misreading circuit. Resistor now reads 10K. This prevents any measurement error

Higher voltage of 1.5 V causes semiconductor to conduct for reading conductivity or front-to-back ratio of junction

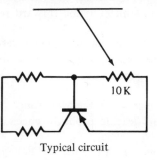

10 K

Typical circuit

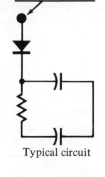

Typical circuit

(a)

(b)

Fig. 1–32   Basic Hi-Lo FETVM ohmmeter functions: (a) Low-voltage ohmmeter function; (b) Conventional ohm-meter function

voltage testing facilities. This type of instrument is often called a Hi-Lo FETVM. Its basic ohmmeter functions are summarized in Fig. 1–32. Because the 0.08-volt test function does not "turn on" semiconductor junctions, more informative and more extensive resistance measurements can be made in solid-state circuitry, including integrated circuits. In other words, all semiconductors are essentially open-circuited at 0.08 volt.

QUESTIONS AND PROBLEMS

True-False

1. The multimeter is the most useful and most basic of all service instruments.
2. The four chief functions of a multimeter are to measure DC voltage, DC current, AC voltage, and AC current.
3. The ammeter is connected in series with a circuit to measure current.
4. Whenever we measure AC voltage that has a DC component, we must use the output function.
5. A multimeter usually has only one scale.
6. There may be several DC voltage scales on a multimeter.
7. The polarity-reversing function is useful when measuring the resistance of a resistor.
8. The reverse-polarity switch is most useful when testing the front-to-back ratio of a diode.

9. The common and +10 A jacks are used when using the 10-A range of the meter shown in Fig. 1–5.

10. The reading of the forward resistance of a diode changes when the range switch of the ohmmeter is changed.

11. The important test of a diode with an ohmmeter is the forward resistance test.

12. The purpose of a multimeter probe is to decrease the voltage range.

13. The high-voltage probe is designed to be used on one range of a voltmeter.

14. The ohms-per-volt rating of a meter is a good indicator of the loading that the meter will produce on a circuit.

15. Signal tracing probes can be used at frequencies in the MHz range.

16. A multimeter can be used with a signal tracing probe to check the neutralizing of a radio transmitter.

17. It is necessary to use the output function to measure the DC voltage on the plate of a vacuum tube.

18. Audio signals are measured in dB.

19. The dB scale is a linear scale

20. Decibel values are power ratios.

21. The dB scale is calibrated with respect to measurements across 1000 ohms.

22. The resistors that carry the extra current in an ammeter are called multipliers.

23. A separate terminal is usually provided for the 10-A range.

24. The meter movement operates on either AC or DC current.

25. The AC function of a multimeter can be used to measure a square wave voltage.

26. The advantage of a TVM over the multimeter is its higher internal resistance on low ranges.

Multiple-Choice
(Choose the Most Correct Answer)

1. The most basic of all test instruments is the
   (a) multimeter.
   (b) oscilloscope.
   (c) ohmmeter.

2. The ammeter is connected in series with a circuit to measure
   (a) voltage.
   (b) resistance.
   (c) current.

3. The voltmeter is connected in parallel with a component to measure
   (a) current.
   (b) voltage.
   (c) resistance.

4. The ohmmeter is used to measure
   (a) current.
   (b) voltage.
   (c) resistance.

5. When an AC voltage with a DC component is to be measured, we
   (a) use the DC scale and subtract.
   (b) use the output function.
   (c) subtract the DC reading from the AC reading.

6. Most multimeters are of the _____ type.
   (a) analog
   (b) digital
   (c) logical

7. Suppose that you are reading a voltage on the 2.5 range and the pointer rests over 225 volts. What is the correct reading?
   (a) 225 volts
   (b) 22.5 volts
   (c) 2.25 volts

8. When the DC voltmeter is connected with the polarity reversed, the most probable effect will be for the
   (a) meter to be damaged.
   (b) meter to show no movement of the pointer.
   (c) pointer to deflect down-scale.

9. The zero ohms control is used to compensate for
   (a) aging of the batteries.
   (b) the leads of the meter.
   (c) the temperature effect.

10. The resistance of a diode is being measured with an ohmmeter on the R×1 range. If the scale of the meter is changed to the R×10 position, the reading will
    (a) increase.
    (b) decrease.
    (c) remain the same.

11. An ohmmeter test will indicate a defective diode by an indication of a
    (a) low front-to-back ratio.
    (b) low forward resistance.
    (c) high reverse resistance.

12. DC probes are usually used with a voltmeter to
    (a) increase the range.
    (b) decrease the range.
    (c) filter the AC signal.

13. A signal tracing probe can be used to check
    (a) low values of DC voltage.
    (b) high values of DC voltage.
    (c) high frequency signals.

14. The output function on a multimeter is used to
    (a) amplify a small signal so that it can be seen on an oscilloscope.
    (b) measure the AC voltage when a DC is present.
    (c) measure the DC voltage when an AC is present.

15. The decibel scale is used to measure the level of
    (a) an AC signal.
    (b) a DC signal.
    (c) an AC current.

16. A separate terminal is usually provided for the 10-A range
    (a) for safety of the operator.
    (b) because it is impractical to pass this current through the range switch.
    (c) for safety of the instrument.
17. The purpose of the rectifier in a multimeter is to
    (a) filter a DC voltage.
    (b) change an AC voltage to a DC voltage.
    (c) protect the meter movement.
18. The AC voltage function of a multimeter will measure
    (a) a sine wave.
    (b) a square wave.
    (c) both a sine wave and a square wave.

General

1. What is the current through a 50-ohm load that has 20 volts dropped across its terminals?
2. How is an ammeter connected to measure current?
3. What function of the multimeter would you use to measure an AC voltage in presence of a DC voltage?
4. How is the voltmeter connected to measure voltage?
5. What is the value of a resistor that indicates 2.5 on the scale when the VOM is on the R×100 scale?
6. Why can the ohmmeter be used to determine the front-to-back ratio of a diode?
7. Between which terminals should you measure a high front-to-back ratio on a transistor?
8. What is the purpose of a high voltage probe?
9. What is the internal resistance of a 20,000 ohms-per-volt meter on the 100-volt range?
10. What is the ohms-per-volt rating of a voltmeter that is built around a 40-microampere meter movement?
11. What are some of the applications of a signal tracing probe?
12. What is the advantage of a TVM over a VOM?
13. Suppose two 1-megohm resistors are connected in series across a 100-volt source. What voltage will a VOM indicate if it has a sensitivity of 20,000 ohms per volt? Assume that the meter is on the 50-volt scale.
14. Does the multimeter usually have a higher useful frequency range than a TVM?
15. Is the input resistance of a TVM the same on every voltage range?
16. Can a TVM be used to measure voltage values in the millivolt range?

2

Oscilloscopes and Vectorscopes

2.1 SURVEY OF OSCILLOSCOPES

Oscilloscopes of various types are used in electronic servicing procedures, and are included in the most useful group of test instruments. Basically, all oscilloscopes are voltmeters, although the additional information provided by oscilloscopes sometimes tends to obscure this fact. That is, any oscilloscope has the function of an AC voltmeter, and many scopes also have the function of a DC voltmeter. The basic oscilloscope is perhaps known best for its ability to display AC waveforms, particularly with respect to time. In other words, most scopes contain a linear *time base*, whereby the rise and fall of a voltage is displayed from one instant to the next. For example, several key television waveforms are shown in Fig. 2–1, with their peak-to-peak voltages.

An oscilloscope has more controls than a multimeter or TVM, as shown in Fig. 2–2. However, most of the controls do not require attention during ordinary servicing procedures. For example, after the focus and astigmatism controls have been adjusted for a sharp trace, they do not require readjustment unless the intensity control is turned up or down. Most service tests are made at a fixed level of the intensity control. These and other details of oscilloscope operation are explained in greater detail subsequently. Note that a DC scope has the same controls as an AC scope. From the application standpoint, an AC scope does not respond to a DC input voltage.

Service-type oscilloscopes are also classified into wide-band and narrow-band types. This terminology denotes the bandwidth of the vertical amplifier, which processes the input signal. Although there is no sharp dividing line, a narrow-band scope has a vertical-amplifier bandwidth of

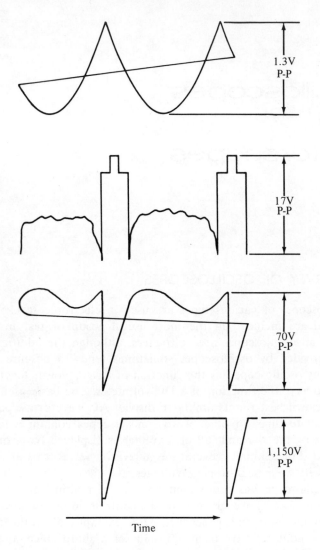

Fig. 2–1   Typical television waveforms and peak-to-peak voltages

3 MHz or less, whereas a wide-band scope has a vertical-amplifier bandwidth of at least 4 MHz. Narrow-band scopes are used primarily in black-and-white TV servicing procedures, whereas wide-band scopes are used also in color TV servicing. To anticipate subsequent discussion, a vertical-amplifier bandwidth of 4 MHz is required to pass the color burst and chroma sidebands, which are the basic components of a color-TV signal.

Another classification is made into triggered-sweep and free-running types. The majority of scopes used in service shops have free-running time

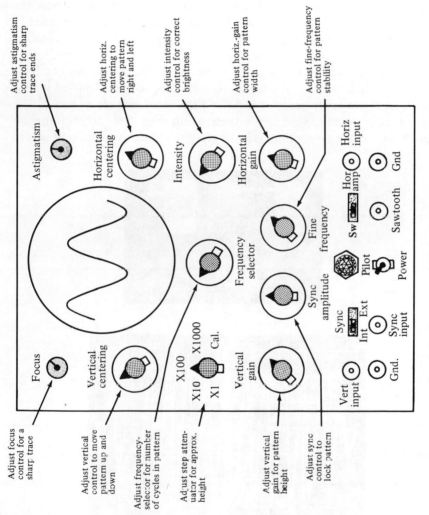

Adjust focus control for a sharp trace

Focus

Adjust vertical control to move pattern up and down

Vertical centering

Adjust frequency-selector for number of cycles in pattern

X100
X10   X1000
X1   Cal.

Adjust step attenuator for approx. height

Vertical gain

Adjust vertical gain for pattern height

Adjust sync control to lock pattern

Adjust astigmatism control for sharp trace ends

Astigmatism

Adjust horiz. centering to move pattern right and left

Horizontal centering

Adjust intensity control for correct brightness

Intensity

Adjust horiz.-gain control for pattern width

Horizontal gain

Adjust fine-frequency control for pattern stability

Frequency selector

Fine frequency

Sync amplitude

Sync
Int Ext

Sync input

Gnd.

Vert input

Hor amp
Sw

Sawtooth

Horiz input

Gnd

Pilot

Power

Fig. 2–2 Operating controls of c basic oscilloscope

39

bases. A free-running or recurrent time base cannot display less than one complete cycle of a signal on the scope screen. On the other hand, a triggered time base can be adjusted to pick out a small section of a waveform, and to expand this section horizontally for evaluation of waveform detail. For example, Fig. 2–3(a) shows a chroma waveform as displayed on the screen of a conventional scope. Next, Fig. 2–3(b) shows how the color burst can be picked out by a triggered-sweep scope and expanded to full screen width. This facility is useful, for example, when troubleshooting and adjusting color-bar generators.

Most oscilloscopes can be used as vectorscopes. Some designs have more utility than others in display of vectorgrams, such as depicted in Fig. 2–4. Basically, a vectorgram is a polar pattern of voltage versus phase, whereas conventional waveforms such as in Fig. 2–1 are rectangular patterns of voltage versus time. The development and evaluation of screen

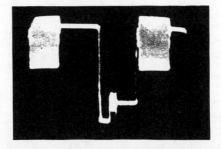

(a)

(b)

Fig. 2–3   Example of triggered-sweep scope application: (a) display of chroma waveform by an ordinary scope; (b) expansion of color burst by triggered-sweep scope

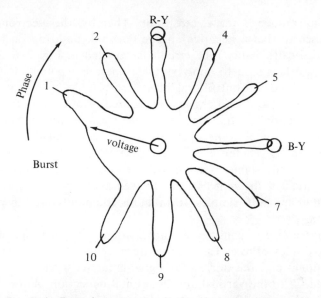

Fig. 2–4   Typical vectorgram displayed on the screen of a vectorscope

patterns will become better understood as subsequent topics are discussed. At this point, we will merely note that vectorgrams are useful in preliminary chroma-troubleshooting procedures because a vectorgram provides a comparatively large amount of circuit-action information in a single pattern.

2.2   CATHODE-RAY TUBE CONSTRUCTION AND OPERATION

As shown in Fig. 2–5, a cathode-ray tube (CRT) contains an electrode arrangement for producing an electron beam. This electron gun is followed by a pair of vertical deflection plates and a pair of horizontal deflection plates. The electron beam can be moved up or down on the screen by application of a voltage to the vertical deflection plates. Similarly, the beam can be moved left or right on the screen by application of a voltage to the horizontal deflection plates. It follows that the beam can be deflected to any point on the screen surface by a suitable combination of vertical and horizontal deflection voltages. When the beam strikes the fluorescent screen, secondary electrons are ejected. These secondary electrons are attracted to the aquadag (graphite) coating on the inner surface of the CRT. From there, the electrons flow to ground and back to the cathode in the electron gun. Thereby, the electron-beam circuit is closed so that a negative charge does not build up on the screen.

A typical cathode-ray tube circuit is depicted in Fig. 2–6. A voltage-divider network is used to apply suitable operating potentials to each of the CRT electrodes. Control-grid bias is adjustable by means of the intensity control, in order to choose the desired level of pattern brightness. Operating potential on the focusing anode is adjustable by means of the focus control, to enable the operator to obtain a spot of minimum size on the screen. DC potentials on the vertical and horizontal deflection plates are also adjustable by means of the centering controls, so that the resting position of the spot can be changed as desired. This arrangement is an oscilloscope in its simplest form and is commonly used in vectorscope application.

It is instructive to consider the application of the arrangement in Fig. 2–6 in the display of basic Lissajous patterns. Note that the vertical signal-input terminal and the horizontal signal-input terminal are AC coupled via C1 and C2 to the vertical and horizontal deflection plates. If a 5-inch CRT is used, from 100 to 150 peak-to-peak volts must be applied to obtain full-screen deflection. The exact sensitivity of the arrangement depends on the type of CRT that is utilized and on the value of high voltage. A high value of accelerating voltage provides maximum pattern brightness, but also makes the electron beam "stiffer" and requires greater deflection voltage.

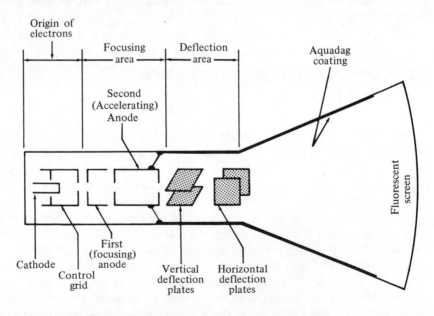

Fig. 2–5   Cross section of a typical cathode-ray tube

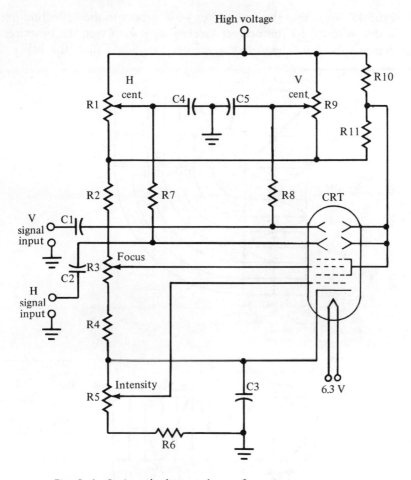

Fig. 2–6 Basic cathode-ray tube configuration

Lissajous patterns are formed by application of sine-wave voltages to the vertical and horizontal deflection plates. A Lissajous pattern provides both frequency and phase information. The simplest examples of Lissajous patterns are depicted in Fig. 2–7 where the vertical and horizontal deflection voltages have the same frequency. In (a) the vertical and horizontal voltages have the same phase, whereas in (b) the vertical and horizontal voltages are 90° out of phase. Note that in the first example, a straight diagonal line is displayed, and that in the second example a circle is displayed. In summary, these are Lissajous patterns which indicate that the two input voltages have the same frequency, and that they have a 0° and a 90° phase relation, respectively.

In addition, we note that both of the Lissajous patterns in Fig. 2–7 are formed by signals that have equal amplitudes. This fact is shown in

(a) by the 45° angle that the line makes with respect to the deflection axes, and is shown in (b) by the perfect circle that is displayed. In practice, it may happen that one signal has a greater amplitude than the other. In

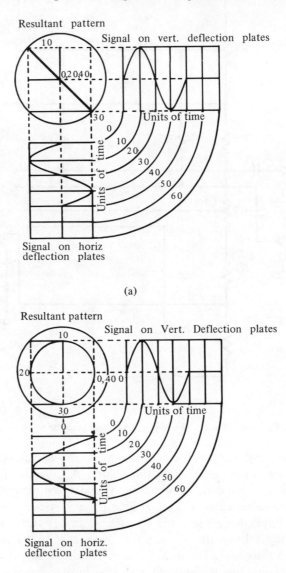

(a)

(b)

Fig. 2–7 Basic Lissajous patterns formed by sine waves: (a) input signals have the same frequency and phase; (b) input signals have the same frequency and a 90° phase difference

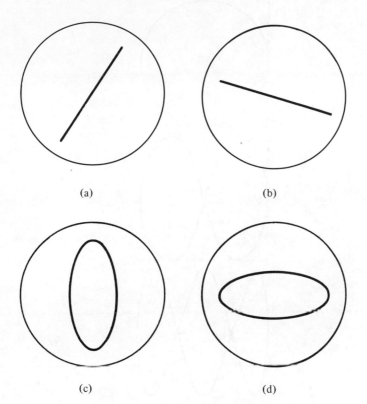

(a)   (b)

(c)   (d)

Fig. 2–8   Lissajous patterns formed by signals with the same
frequency but with different amplitudes: (a) input
signals in phase, vertical signal greater than horizontal signal; (b) input signals in phase, horizontal
signal greater than vertical signal; (c) input signals
90° out of phase, vertical signal greater than horizontal signal; (d) input signals 90° out of phase,
horizontal signal greater than vertical signal

such a case, we observe a line pattern with an angle other than 45°, or we
observe an ellipse instead of a circle in case of a 90° phase difference
between the signals. These patterns are depicted in Fig. 2–8. Note that
the major axes of the ellipses are either vertical or horizontal. This fact
indicates that there is a 90° phase difference between the two input signals,
although their amplitudes are different.

Next, let us consider the Lissajous patterns that are formed when the
input signals have the same amplitude, are 90° out of phase, and have
frequencies that are integrally related. In this situation, basic patterns are
formed as in Fig. 2–9. Such patterns are commonly used in calibration of
test equipment such as audio oscillators. We will find that a 1-to-1 pattern,
for example, will "stand still" on the CRT screen only when the input

Patterns   Freq. ratio

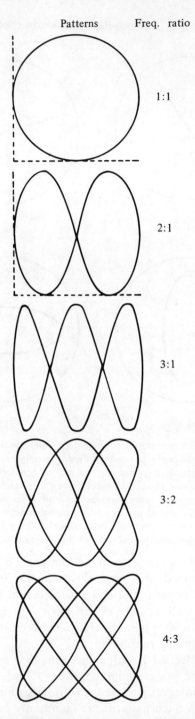

1:1

2:1

3:1

3:2

4:3

Fig. 2–9   Basic Lissajous patterns for signals that are 90°
out of phase, with the same amplitudes and with
integrally related frequencies

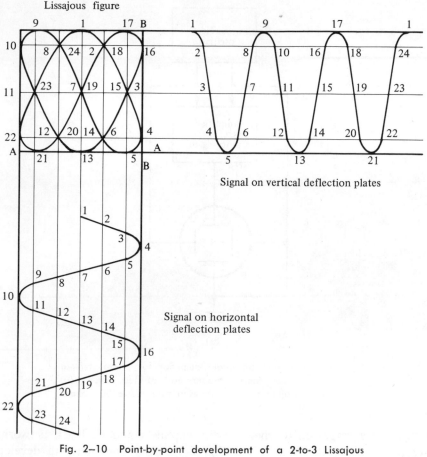

Fig. 2–10   Point-by-point development of a 2-to-3 Lissajous
pattern

signals have exactly the same frequency. Otherwise, the pattern "writhes"
on the screen and continually changes shape. A stable pattern is next
obtained when the input signals have a 2-to-1 frequency ratio, and so on.

It is important to observe that the frequency ratio in a Lissajous pattern
is given by the tangencies to the vertical and horizontal axes. For example,
we note in Fig. 2–9 that the 1 to 1 frequency ratio produces a circle which
has a single point of tangency vertically as well as horizontally. Next,
the 2-to-1 frequency ratio produces a bow-tie or lazy-8 pattern which has
a single point of tangency vertically and two points of tangency horizontally. Thus, the operator can easily determine what harmonic frequency
he may be employing by observing the number of tangencies. The point-
by-point development of a 2-to-3 Lissajous pattern is depicted in Fig. 2–10.

As noted previously, a Lissajous pattern also provides phase information, which is very useful in servicing procedures. For example, Lissajous

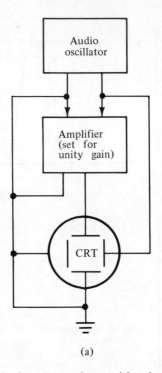

(a)

Fig. 2–11   Checking an audio amplifier for phase shift at various input frequencies: (a) test setup; (b) Lissajous patterns showing progressive 30° phase shifts

patterns are employed to check audio amplifiers and feedback networks for phase shift, as illustrated in Fig. 2–11. Any amplifier will develop phase shift near its low- and high-frequency limits, although a normally operating amplifier will have negligible phase shift over its rated frequency range, such as from 20 Hz to 20 kHz. We observe in Fig. 2–11(b) that progressive phase shifts cause a diagonal straight-line pattern to change into ellipses of lessening eccentricities until a circular pattern is formed at 90° phase difference. Thereafter, additional phase shifts cause ellipses of opposite inclination and increasing eccentricities to be displayed, until finally at 180° phase difference a straight diagonal line of opposite inclination is displayed.

A simple calculation permits the scope operator to evaluate the exact amount of phase difference in any elliptical Lissajous pattern, as shown in Fig. 2–12. The pattern is first centered on the CRT screen, and then the vertical intervals M and N are measured. The ratio of M/N is equal to the sine of the phase-difference angle. Therefore, by consulting a table of sines in any trigonometry book, the phase angle can be easily found. Of

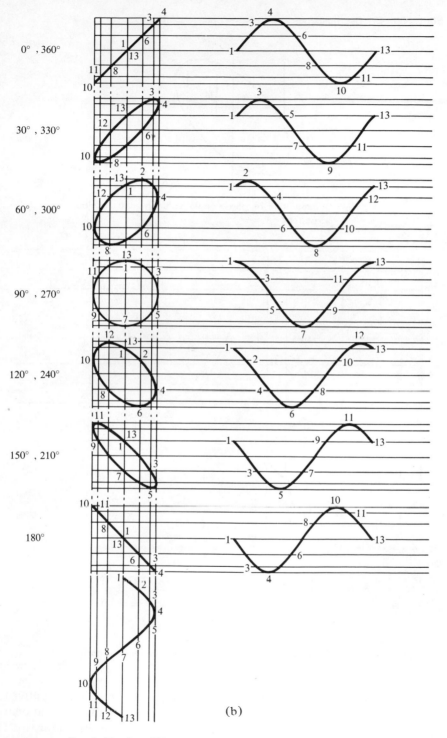

0° , 360°

30° , 330°

60° , 300°

90° , 270°

120° , 240°

150° , 210°

180°

(b)

Fig. 2–11   (cont'd.)

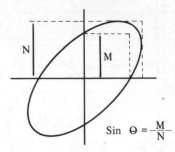

$$\text{Sin } \Theta = \frac{M}{N}$$

Fig. 2–12 Evaluation of phase-angle value from a Lissajous pattern

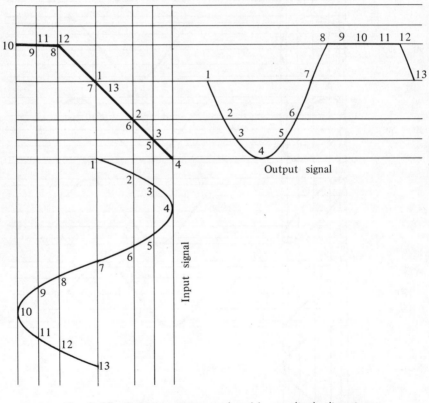

Fig. 2–13 Lissajous pattern produced by amplitude distortion in an amplifier

course, if the phase difference is greater than 90°, the elliptical pattern will have an opposite inclination, and we make use of the conventional trigonometric formulas to find the angle. For example, if we find a value

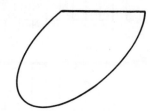

Fig. 2–14 Lissajous pattern resulting from clipping accompanied by phase shift

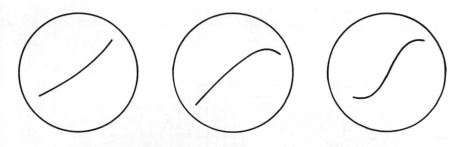

Fig. 2–15 Lissajous patterns produced by various amounts of signal compression in an amplifier

of 60° for the phase angle in the table of sines, but the ellipse has an opposite inclination, we subtract 60° from 360° to find the actual angle, viz., 300°.

Lissajous patterns are also used to determine whether an audio amplifier is distorting the signal, as shown in Fig. 2–13. In this example, the output signal is clipped on its positive peak. In turn, the Lissajous pattern is not a straight line, but suddenly changes direction at the clipping point. In case the clipping action is accompanied by phase shift, the Lissajous pattern is displayed as a flattened ellipse, as exemplified in Fig. 2–14. In case amplitude distortion is taking place as signal compression, instead of signal clipping, a curved diagonal line pattern is displayed, as shown in Fig. 2–15. Again, if signal compression is accompanied by phase shift, an egg-shaped ellipse is displayed on the CRT screen.

Next, it is instructive to consider the basic Lissajous patterns that are called vectorgrams and are employed in color-TV servicing procedures. Again, the CRT arrangement shown in Fig. 2–6 is most often employed as a vectorscope. The chroma section of a color-TV receiver develops R-Y and B-Y output waveforms which are depicted in Fig. 2–16 in idealized form. It is helpful to start our analysis of vectorgrams with the waveforms that would be produced if circuit action were perfect. When these R-Y and B-Y waveforms are applied to the vertical and horizontal

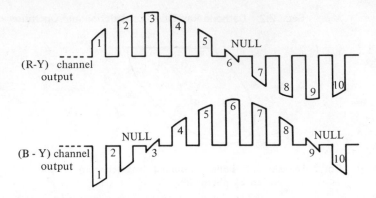

Fig. 2–16 Idealized R-Y and B-Y output waveforms from the chroma section of a color-TV receiver

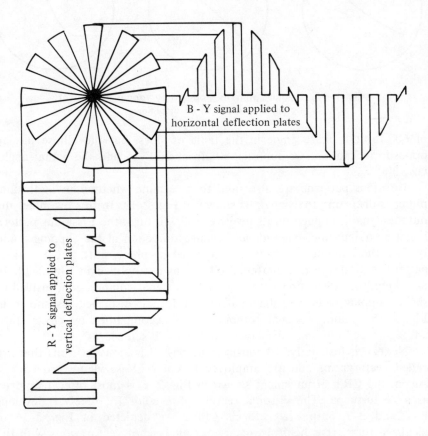

Fig. 2–17 Vectorgram produced by ideal R-Y and B-Y wave-forms

deflection plates respectively in a vectorscope, the vectorgram shown in Fig. 2–17 is displayed on the screen. That is, 12 "petals" are formed, radiating from the center of the Lissajous pattern.

In case the bandwidth of the R-Y and B-Y demodulator circuits is less than normal, a characteristic distortion of the output waveforms occurs in which the baselines are not straight, but curved. The effect of this baseline curvature is depicted in Fig. 2–18. We note that the "petals" of the vectorgram do not start at the center of the pattern in this case. Instead, the petals start around the circumference of an inner circle. The diameter of this inner circle corresponds to the amount of bandwidth limitation. A practical example of bandwidth indication is seen in Fig. 2–4. Note in passing that most color-TV receivers produce vectorgrams that have at least a small inner circle. The receiver service data is the final authority in this matter.

Next, let us consider the screen display produced by a vectorscope in case the R-Y channel is dead, or the B-Y channel is dead. As shown in

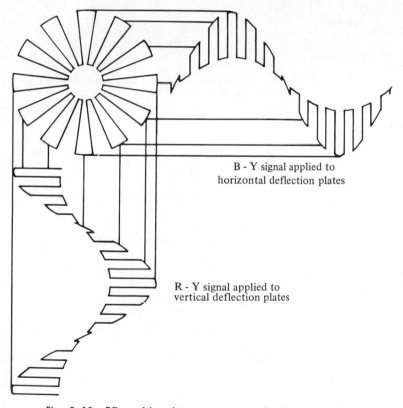

B - Y signal applied to horizontal deflection plates

R - Y signal applied to vertical deflection plates

Fig. 2–18   Effect of baseline curvature on the formation of a vectorgram

Fig. 2–19, a dead R-Y channel results in the display of a horizontal line, and a dead B-Y channel results in the display of a vertical line. In practice, the normal amplitudes of the R-Y and B-Y signals are not exactly equal, and in some receivers they may be substantially unequal. As would be expected, if the R-Y signal has a greater amplitude than the B-Y signal, an elliptical vectorgram is formed, as depicted in Fig. 2–20. Note also that most receivers employ a horizontal-blanking interval, with the

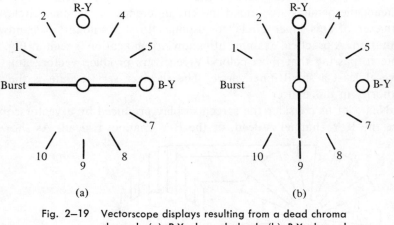

Fig. 2–19   Vectorscope displays resulting from a dead chroma channel: (a) R-Y channel dead; (b) B-Y channel dead

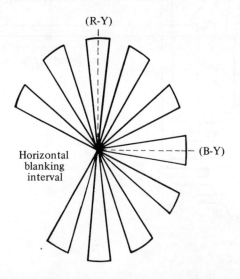

Fig. 2–20   Idealized 10-petal vectorgram, with the R-Y signal at higher amplitude than the B-Y signal

result that the vectorgram contains 10 "petals." A few receivers will display 11 "petals," but we never observe 12 "petals" in practice. The reason for this is seen in Fig. 2–21. There is no chroma signal present during the horizontal sync-pulse interval.

It is instructive to consider some of the basic facts of circuit action which cause practical vectorgrams to depart from the ideal with respect to petal shapes. First, limited receiver bandwidth results in rounded corners instead of sharp corners in chroma waveforms, as presented in Fig. 2–22(a). In turn, the vectorgram petals display rounded corners, as shown in Fig. 2–22(b). Second, residual phase errors in the chroma-demodulator circuits result in output waveforms that do not have quite the same shapes. For example, Fig. 2–23 exemplifies R-Y and B-Y waveforms that do not have identical waveshapes, with the result that the vectorgram petal waveshape is affected accordingly.

Early color-TV receivers employed quadrature demodulation on the R-Y and B-Y axes. However, various departures from quadrature demodulation are utilized in later receivers. One common variation is the XZ demodulation system in which the axes are 105° apart. In turn, a normally operating XZ demodulator has the somewhat idealized output waveforms depicted in Fig. 2–24(a). From the viewpoint of Lissajous-pattern formation in a CRT, this situation introduces an effective 15° phase shift into the vectorgram, because the vertical and horizontal deflection plates are located 90° apart. A practical vectorgram displayed when X and Z signals are applied to a vectorscope is shown in Fig. 2–24(b). Since pattern analysis becomes somewhat involved in this situation, TV technicians generally make comparison tests only and do not attempt detailed evaluation of pattern distortions that may occur.

It follows from previous discussion that one of the important facts obtained from vectorscope servicing tests is the determination of the chroma-demodulation phase angle. For example, Fig. 2–25 shows the phase-shift circuits used in a typical R-Y/B-Y/G-Y demodulation system. When the R-Y and B-Y demodulator output signals are used to display a vectorgram, the major and minor axes of the elliptical vectorgram should be vertical and horizontal. If the axes are inclined, it is indicated that there is a defective component in the R-Y and B-Y phase-shifting networks. Capacitors are the most common troublemakers. Note that the relative amplitudes of the R-Y and B-Y signals are of no consequence in phase evaluation. For example, Fig. 2–26 depicts five elliptical patterns with varying B-Y amplitudes. In each case, a 90° demodulation angle is indicated because the major and minor axes are vertical and horizontal.

Another important fact provided by a vectorgram is the amplitude linearity or nonlinearity of the chroma channel. For example, Fig. 2–27

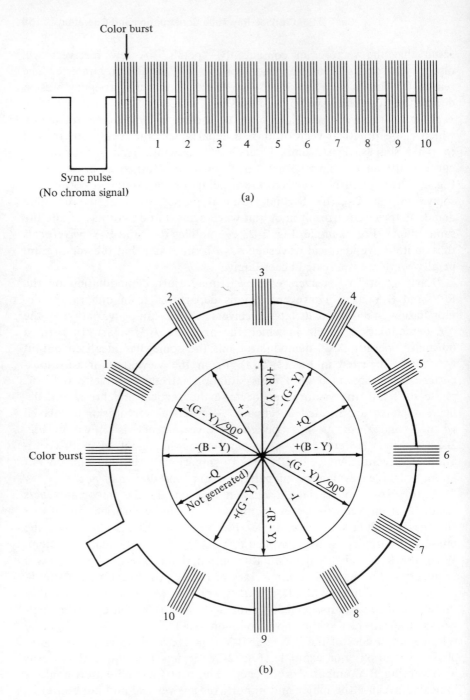

Fig. 2–21   Relation of a keyed-rainbow signal to vectorgram
formation: (a) signal waveform; (b) signal phases
in vectorgram pattern

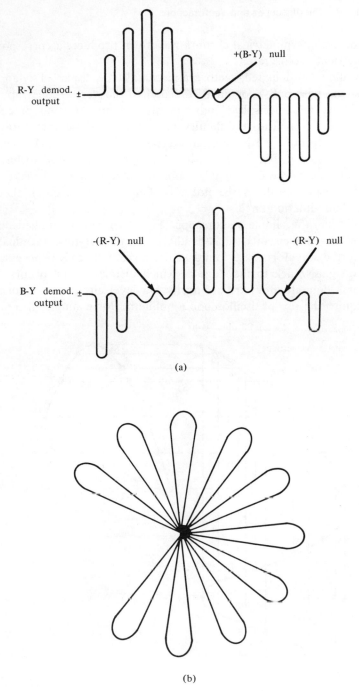

R-Y demod.
output ±

+(B-Y) null

-(R-Y) null

-(R-Y) null

B-Y demod. ±
output

(a)

(b)

Fig. 2–22 Effect of rounded corners on vectorgram pattern:
(a) rounded corners on chroma output waveforms;
(b) rounded corners on vectorgram petals

shows how an unsymmetrical chroma waveform produces a corresponding unsymmetrical vectorgram.

As would be anticipated, vectorgrams can also be displayed by applying a G-Y signal with a B-Y or an R-Y signal to the vectorscope input terminals. However, this is less common practice in servicing procedures because of the comparative difficulty in evaluating these vectorgrams. It is instructive to note the basic characteristics of a G-Y and B-Y vectorgram, as depicted in Fig. 2–28. If the G-Y signal is adjusted to have the same amplitude as the B-Y signal, a normal vectorgram will make a 45° angle with respect to the horizontal axis. This provides a quick check on the G-Y demodulation angle.

In practice, the simple vectorscope (Fig. 2–6) can be applied only in high-level chroma circuits, such as the color picture-tube terminals. In case distorted waveforms are found at the picture tube, it is necessary to trace the signals back to the stage at which distortion first occurs. Since the chroma demodulators are ordinarily low-level circuits, waveform displays require the use of oscilloscope amplifiers. For example, if a vector-

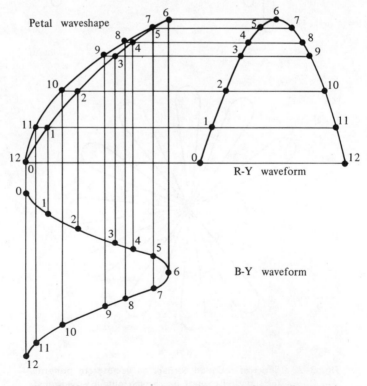

Fig. 2–23 Differences in R-Y and B-Y waveforms affect petal shape in the vectorgram

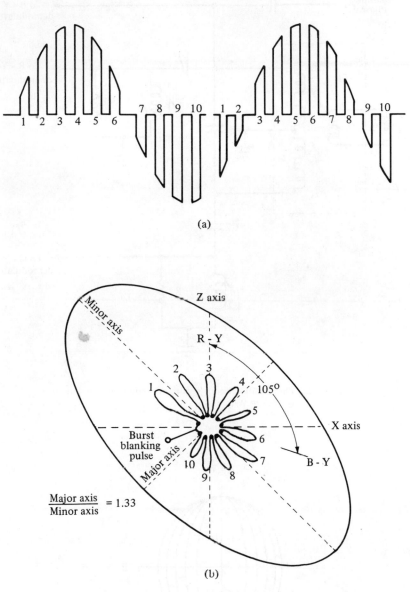

(a)

(b)

Fig. 2–24 Vectorgrams may involve demodulation angles
other than 90°: (a) idealized output waveforms
from 105° XZ system; (b) practical vectorgram
pattern for an XZ system

gram is to be displayed directly from the output signals of the R-Y and
B-Y demodulators, both the vertical and horizontal amplifiers of the scope
must be utilized, as depicted in Fig. 2-29. It must be stressed that only

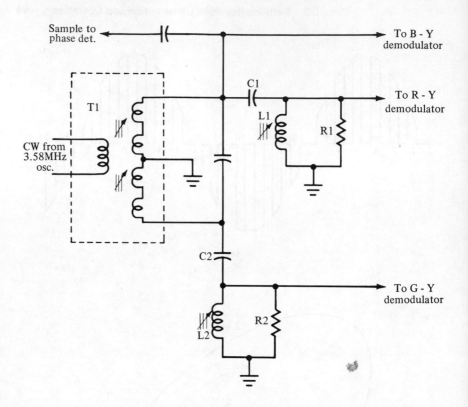

Fig. 2–25 Phase-shift circuitry for demodulator section of an R-Y/B-Y/G-Y system

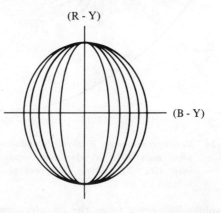

Fig. 2–26 Variation in pattern aspect due to different relative amplitudes of the R-Y and B-Y signals

the better-quality service scopes are suitable for this type of test work. In other words, the horizontal amplifier of most service-type scopes will distort a chroma waveform seriously, and the vertical amplifier may introduce appreciable distortion.

From a technical standpoint, the bandwidth of the arrangement in Fig. 2–6 greatly exceeds the bandwidth of any service-type scope amplifier. The phase error and nonlinearity of the CRT is also negligible. In other words, limitations in oscilloscope performance are not imposed by the CRT, but by associated electronic sections, such as the vertical and horizontal amplifiers. These considerations are explained in greater detail in the following chapter. When a complete oscilloscope is used as a vectorscope, with the input signals coupled directly to the CRT deflection plates, the

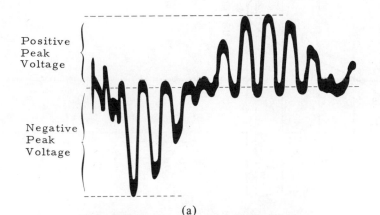

Positive
Peak
Voltage

Negative
Peak
Voltage

(a)

(b)

Fig. 2–27   Amplitude nonlinearity in chroma waveforms: (a) unsymmetrical chroma-demodulator output waveform; (b) unsymmetrical vectorgram

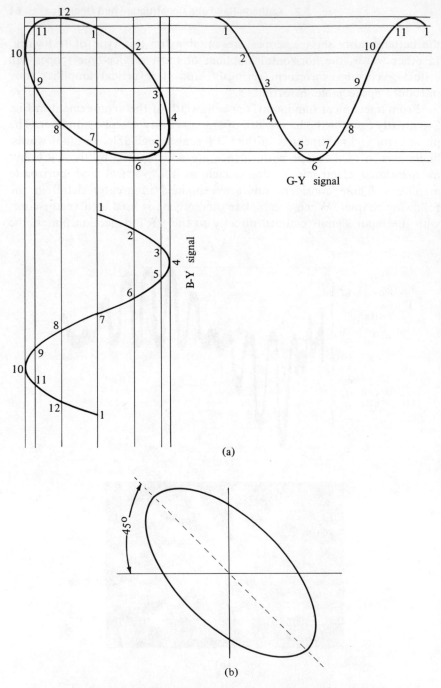

G-Y signal

B-Y signal

(a)

45°

(b)

Fig. 2–28 Development of a vectorgram from G-Y and B-Y signals: (a) point-by-point correspondences; (b) major axis normally inclined at 45°.

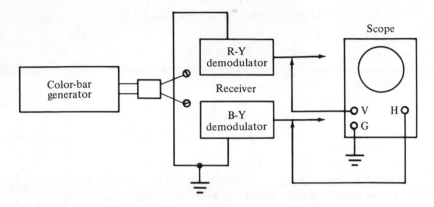

Fig. 2–29   Vectorscope input signals may be stepped up
through vertical and horizontal scope amplifiers

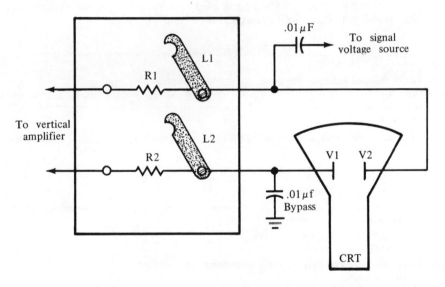

Fig. 2–30   Links L1 and L2 are opened to effectively isolate
the deflection plates from the scope amplifiers

deflection plates must be effectively isolated from the vertical and horizontal amplifiers. For this reason, it is customary to provide connecting
links behind the CRT, as depicted in Fig. 2–30. Resistors R1 and R2
have a high value, such as 2 megohms, and are left in the circuit so that
the centering controls remain operative. The input signal is then capacitively coupled to one deflection plate, while the opposing deflection plate
is bypassed to ground.

QUESTIONS AND PROBLEMS

True-False

1. Oscilloscopes are basically voltmeters.
2. What is the function of AC coupling?
3. The oscilloscope has more controls than a multimeter.
4. A free-running oscilloscope cannot be used to display less than one cycle of a waveform.
5. Most oscilloscopes can be used as vectorscopes.
6. Lissajous patterns are used to measure voltage of a waveform.
7. Lissajous patterns can be used to measure phase differences.
8. A vectorgram is a type of Lissajous pattern.
9. When a service-type oscilloscope is used as a vectorscope, the horizontal amplifier is set to maximum gain.
10. When a service-type oscilloscope is used as a vectorscope, the vertical and horizontal amplifiers are bypassed and isolated.

Multiple-Choice

1. Oscilloscopes indicate the _____ value of a voltage.
   (a) DC
   (b) rms AC
   (c) peak-to-peak
2. The term wide-band denotes bandwidth of the _____ amplifier(s) of an oscilloscope.
   (a) vertical
   (b) horizontal
   (c) vertical and horizontal
3. We must use a _____ to observe a portion of a waveform.
   (a) vectorscope
   (b) triggered-sweep oscilloscope
   (c) free-running oscilloscope
4. Lissajous patterns can be used to determine
   (a) phase shift.
   (b) voltage amplitude.
   (c) amplitude distortion.
5. When a service type oscilloscope is used as a vectorscope,
   (a) the vertical amplifier is turned off.
   (b) the horizontal amplifier is turned off.
   (c) the vertical and horizontal amplifiers are isolated.

General

1. What are the primary purposes of an oscilloscope?
2. How is the sweep generator triggered in a triggered oscilloscope?

3. How is an oscilloscope used as a vectorscope?
4. Suppose two signals 90° out of phase are applied to the vertical and the horizontal inputs of the oscilloscope. What is the pattern on the CRT?
5. Suppose you are measuring the phase with Lissajous patterns and N is 2 units and M is 4 units. What is the phase angle?
6. What are the requirements for an oscilloscope to be used as a vectorscope?
7. What is the purpose of the connecting links behind the CRT?

3

Oscilloscope
Operation and
Application

3.1 OPERATIONAL SECTIONS OF BASIC
OSCILLOSCOPES

Service-type oscilloscopes are provided with the functional sections shown in Fig. 3–1. The vertical amplifier enables tests in low-level circuits. For example, a typical vertical amplifier provides a sensitivity of 0.02 volt per inch on the CRT screen. Since the CRT has a sensitivity of approximately 30 volts per inch, it follows that the vertical amplifier has a voltage gain of 1500 times. Note that the time base generates a sawtooth output waveform which provides horizontal deflection of the electron beam. This type of time base produces deflection that is linear in time. A horizontal amplifier is included to step up the amplitude of the sawtooth voltage waveform sufficiently for full-screen deflection. A voltage gain of 50 times is typical.

Note that the time-base oscillator in Fig. 3–1 is free-running. This term means that the time base operates continuously, whether an input signal is applied or not. However, the time base is not completely independent of an input signal. When an input signal is applied, the sync system feeds a sample of the signal to the time-base oscillator. In turn, the oscillator frequency falls in step with that of the sync signal. This synchronizing action is required to maintain the pattern in a fixed position on the CRT screen. The power supply has low- and high-voltage outputs. Low supply voltages are required by transistors or other active devices, whereas a high supply voltage is required for operation of the CRT. For example, a 5-inch CRT may operate from a 1500 or 2000 volt supply. However, the current demand is less than 1 milliampere.

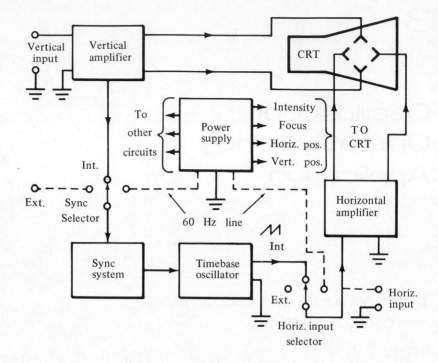

Fig. 3—1 Functional sections of a service-type oscilloscope

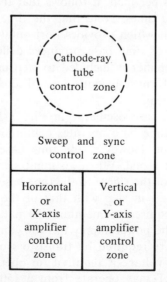

Fig. 3—2 Control layout for a typical service-type oscilloscope

3.2 OSCILLOSCOPE CONTROL FUNCTIONS

Control layouts vary somewhat from one oscilloscope to another. However, the arrangement shown in Fig. 3–2 is typical for a service-type scope. The CRT control zone usually comprises intensity, focus, vertical centering, and horizontal centering controls, as shown in Fig. 3–3. If the intensity control is turned completely counterclockwise, the screen becomes dark. On the other hand, if the intensity control is turned too far clockwise, the pattern becomes excessively bright, and the screen phosphor may be burned. The focus control is adjusted to obtain a sharply defined and thin trace. Since a centered pattern is usually employed, the centering controls are adjusted accordingly.

We note in Fig. 3–3 that the sweep and sync control zone comprises a step-type (coarse) sweep selector, a continuous (fine) sawtooth frequency, and a sawtooth synchronizing control. We set the step control to obtain a display of several complete cycles of the signal, and then adjust the fine control to obtain a lesser number of cycles in the display. Of course, it is impractical to display less than one complete cycle with a free-running sweep oscillator. That is, if an excessive sweep speed is employed, a broken-up and overlapping pattern is displayed. Next, we will find that the sync control must be advanced to a point such that the pattern is locked in a fixed position on the screen. On the other hand, if the sync control is advanced too far, false triggering action occurs, with the result that the pattern becomes distorted and broken.

As shown in Fig. 3–3, most oscilloscopes provide a step-type (coarse) vertical gain control and a continuous (fine) vertical gain control. We use the fine control to "fill in" between the steps of the coarse control. In most applications, the pattern height is adjusted to approximately ⅔ of full screen. Most service scopes have a single horizontal gain control of the continuous type. This control is usually adjusted so that the pattern is expanded horizontally to almost full-screen. To inspect waveform detail, a pattern may occasionally be expanded off-screen both horizontally and vertically, and the interval of interest brought into view by adjustment of the centering controls.

In addition to intensity and focus controls, some oscilloscopes provide an astigmatism control also. This control serves to obtain optimum focus over the entire screen area. For example, when a sine-wave pattern occupies most of the screen, the focus control must be set to somewhat different positions in order to bring the center of the pattern into focus, or to bring the edges of the pattern into focus. In this situation, the astigmatism control is adjusted to obtain uniform focus of the entire waveform. The focus control and the astigmatism control tend to interact.

CRT control zone

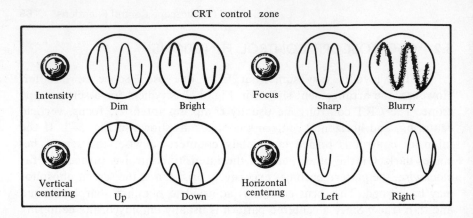

Sweep and sync control zone

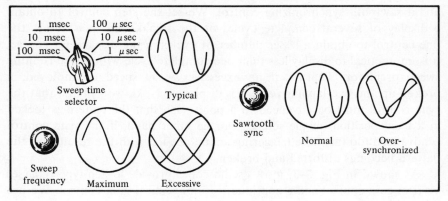

Vertical control zone Horizontal control zone

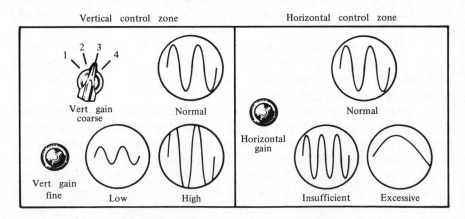

Fig. 3—3 Illustration of oscilloscope control functions

3.3 TIME-BASE ACTION

A linear time base operates by means of a sawtooth deflection voltage on the horizontal deflection plates, as depicted in Fig. 3–4. In this example, the repetition rate of the sawtooth waveform is the same as that of the signal voltage on the vertical deflection plates. Since the sawtooth wave has a small flyback or retrace time (from instant 9 to instant 10), a small portion of the signal is lost on retrace. Although the retrace time can be minimized by suitable design of the sawtooth oscillator, it cannot be reduced to zero. Accordingly, there is always a residual loss of signal information on retrace.

Next, let us observe the relations shown in Fig. 3–5. In this example the repetition rate of the sawtooth waveform is one-third that of the input signal voltage. Accordingly, two complete cycles and most of a third cycle are displayed on the CRT screen.

We will find that the retrace excursion may be visible in a screen pattern, as exemplified in Fig. 3–6. However, most service-type scopes have retrace blanking action provided, whereby the retrace excursion is made invisible. (See Fig. 3–7.) Retrace blanking takes place only when sawtooth

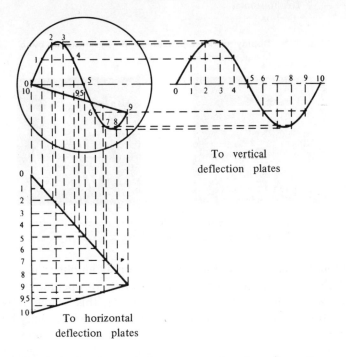

To vertical
deflection plates

To horizontal
deflection plates

Fig. 3–4 Development of a screen pattern by a linear time base

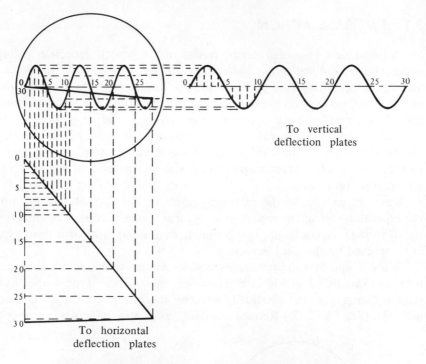

To vertical
deflection plates

To horizontal
deflection plates

Fig. 3–5 Display of three cycles of the input signal by a
linear time base

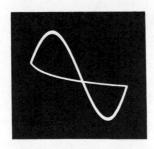 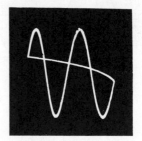

Fig. 3–6 Screen patterns with visible retrace excursions
(Courtesy of U.S. Armed Forces)

sweep is operative. For example, the horizontal amplifier in Fig. 3–1 is provided with an "Int-Ext" switch. In this mode of operation, there is no sawtooth sweep and no blanking action. When this switch is set to its "Ext" position, a sine-wave or other special sweep voltage may be applied from a generator or equivalent source. The most common application for external sweep is in visual alignment procedures. Some service-type scopes have a built-in 60-Hz deflection function, but most require an external 60-Hz source.

Figure 3–8 shows the most common 60-Hz horizontal deflection arrangement, in which the horizontal-input voltage is obtained from a sweep-frequency generator. Although horizontal sawtooth deflection could be employed in this application, optimum pattern linearity is obtained by utilizing 60-Hz sine-wave deflection. The reason for this requirement is that service-type sweep-frequency generators are usually designed with 60-Hz sine-wave frequency modulators. In turn, 60-Hz sine-wave deflection

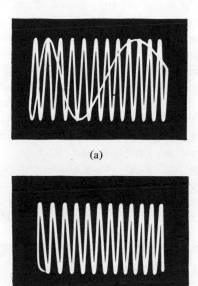

(a)

(b)

Fig. 3–7   Pattern aspect with and without retrace blanking:
(a) retrace unblanked; (b) blanking pulse applied
to CRT grid (Courtesy of U.S. Armed Forces)

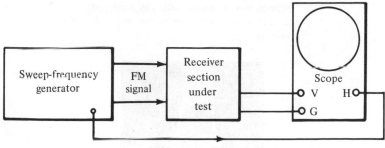

Fig. 3–8   Application of 60-Hz horizontal deflection voltage
to scope from a sweep-frequency generator

is required to obtain optimum horizontal pattern linearity. Insofar as retrace is concerned, we then have the situation illustrated in Fig. 3–9. A frequency-response curve is displayed both on the positive half-cycle and on the negative half-cycle of horizontal deflection.

There are two methods of coping with this situation. Both methods of retrace processing require that the 60-Hz sine-wave deflection voltage be brought into phase with the FM modulator voltage. This is accomplished by adjustment of a phasing control, provided on the sweep-frequency generator. Or, if the scope has a built-in 60-Hz sine-wave deflection facility, a phasing control is included. When the horizontal phasing control is correctly adjusted, the pattern shown in Fig. 3–9(a) is displayed as seen in (b). In this manner, "layover" of trace and retrace is obtained. Sometimes the pattern is utilized in this form. However, the disadvantage of this "layover" pattern is that the forward trace and reverse trace seldom coincide exactly, due to residual system reactances and waveform distortions. Therefore, most technicians prefer to utilize a retrace blanking method that eliminates any double-image effect.

Retrace blanking action in sweep-alignment procedures is usually accomplished in the sweep-frequency generator, but can also be effected in the oscilloscope. When a generator provides a retrace-blanking function, a blanking switch is mounted on the control panel. If the blanking switch

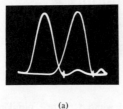
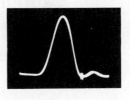

(a)   (b)

Fig. 3–9   Sweep-frequency response curves displayed on 60-Hz sine-wave horizontal deflection: (a) deflection voltage out of phase with FM modulator voltage; (b) deflection voltage in phase with FM modulator voltage

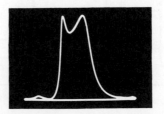

Fig. 3–10   Retrace excursion changed into a horizontal base line

is turned ON, the reverse trace in a frequency-response curve is changed into a horizontal base line, as illustrated in Fig. 3–10. This is accomplished in the generator by gating the FM sweep oscillator off during the reverse-trace interval. On the other hand, if a reverse-trace blanking function is provided in an oscilloscope, a suitably phased 60-Hz bias voltage is applied to the control grid in the CRT. In turn, the CRT beam is cut off during the negative excursion of this blanking bias voltage. The phasing of the bias voltage is adjustable by means of a phasing control located on the control panel of the scope.

3.4 VERTICAL AMPLIFIER CALIBRATION

As previously noted, a scope is basically a voltmeter. It is generally used to measure peak-to-peak values, since these are specified in TV receiver service data. If a scope has DC response, it can be conveniently calibrated from a DC source, as depicted in Fig. 3–11. For example, if a DC scope is to be calibrated for 1 peak-to-peak volt per inch sensitivity, we can apply a 1.5-volt DC source, and adjust the vertical gain control to raise or lower the trace 1.5 inches on the screen. In other words, a DC calibration is numerically the same as a peak-to-peak AC calibration. Note in passing that when a scope has DC response, it is provided with a switch on the control panel so that it can be operated in the AC mode. If the switch is set to its AC position, it is evident that the response depicted in Fig. 3–11 cannot be obtained.

Next, let us consider the calibration of a scope from an AC voltage source. For example, if we employ a 1-volt rms sine-wave voltage, as depicted in Fig. 3–12, its peak-to-peak value is 2.83 volts p-p. Therefore, when this calibrating voltage is applied to the vertical input terminals of a scope, we may calibrate the response for 1 peak-to-peak volt per inch by adjusting the vertical-gain control for 2.83 inches of deflection. In many service shops, a 6.3-volt rms sine-wave source is used for scope calibration. In this case, the peak-to-peak calibrating voltage is 17.8 volts p-p. Note that quite a few service-type scopes have a built-in 1 volt p-p calibrating source. This voltage may be available at a terminal on the control panel, or it may be automatically applied to the vertical input terminals when the vertical step attenuator is switched to "Cal." position.

It is important to note that most AC meters indicate the rms values of sine waves. Since a scope is calibrated in terms of peak-to-peak values, we must not forget to convert an rms meter reading into its peak-to-peak equivalent before proceeding with calibration procedures. Note also that the conversion factor depicted in Fig. 3–12 (p-p equals 2.83 times rms) is true only for a sine wave. It is approximately true for slightly distorted sine waves, and is much in error for badly distorted sine waves. Therefore,

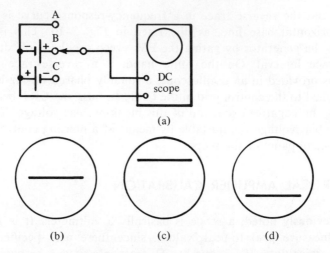

Fig. 3–11   Calibration of a DC scope from a battery voltage source: (a) test setup; (b) zero volts applied; (c) positive volts applied; (d) negative volts applied (Courtesy of Allied Radio Corp.)

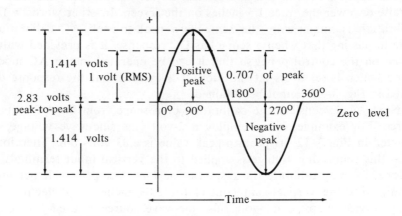

Fig. 3–12   Voltage specifications of a sine wave

the calibrating source should provide a good sine waveform. On the other hand, if we measure the source voltage in peak-to-peak values with a TVM, the waveform is of no concern. That is, a peak-to-peak voltage reading is true regardless of the voltage waveform.

3.5   VERTICAL AMPLIFIER ATTENUATORS

An oscilloscope utilized in TV servicing procedures requires a high input impedance to minimize circuit loading. If the circuit under test is loaded appreciably, the signal voltage decreases and complex waveforms such as sync pulses become distorted. In some cases, circuit loading disturbs the circuit action to such an extent that the section appears to be "dead." There are two components to the input impedance of a scope. Thus, we are concerned with the value of input resistance and with the value of shunt capacitance. In the first analysis, high input resistance can be obtained by employing a 1-megohm potentiometer as the vertical attenuator. Low input capacitance can be obtained by operating the scope with open test leads. However, there are disadvantages to this simple vertical-input system.

When open test leads are used, as depicted in Fig. 3–13(a), stray field pickup becomes an annoying problem. If we are testing in a low-impedance circuit, stray field pickup becomes negligible. On the other hand, when we are testing in a high-impedance circuit, 60-Hz hum voltages produce excessive interference in the pattern. Or, if we are testing near the horizontal sweep section of a TV receiver, flyback pulses produce serious interference, even when testing in medium-impedance circuits. Therefore, it is necessary to use a shielded input cable to the vertical amplifier, as shown in Fig. 3–13(b). This type of input lead eliminates trouble from stray field pickup. On the other hand, the input capacitance to the shielded cable is objectionably high (typically 75 pF), and the action of many TV circuits will be disturbed by this capacitive loading. Therefore, an elaborate type of input system is required for TV servicing applications.

Even when testing in very low impedance circuits, we will find another disadvantage in the arrangement of Fig. 3–13(b). That is, the full frequency response of the vertical amplifier can be obtained only when the

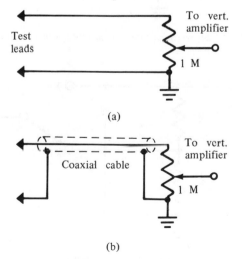

Fig. 3–13   Simple scope input arrangements: (a) open test leads; (b) shielded input cable

potentiometer is set for maximum gain. As we proceed to lower settings on the potentiometer, the frequency response becomes poorer and serious waveform distortion occurs. The reason for this difficulty is seen in Fig. 3–14. Note that the arm of the potentiometer divides the resistance element into two sections, R1 and R2. There is a certain amount of stray capacitance C1 effectively in shunt to R1, and another amount of stray capacitance C2 in shunt to R2. These values of stray capacitance vary from one setting of the potentiometer to another. In turn, the impedances of the R1 and R2 sections change in some arbitrary manner. An RC network of this type will always produce frequency distortion except for one special condition, which we note next.

With reference to Fig. 3–15, a basic step-attenuator circuit is shown, comprising resistors R1 and R2, shunted respectively by trimmer capacitors C1 and C2. These are maintenance adjustments and are used in operating the scope. Note carefully that when the time constants of the two attenuator sections are equal, no frequency distortion takes place. In other words, if R1C1 equals R2C2, it can be shown mathematically

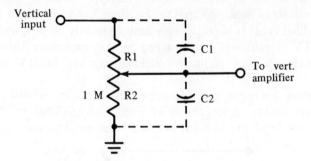

Fig. 3–14   Stray capacitances associated with a simple potentiometer

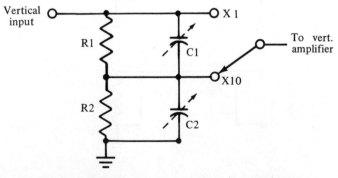

Fig. 3–15   Basic plan of a compensated vertical-input attenuator

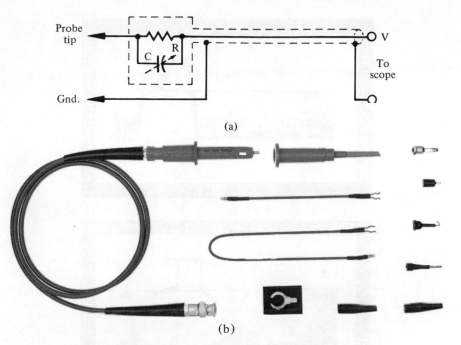

Fig. 3–16   Low-capacitance probe for vertical input: (a) basic
configuration; (b) appearance of a typical low-C
probe (Courtesy of B & K Mfg. Co.)

and experimentally that the divider network is compensated and introduces no frequency distortion. In this example, two steps are provided,
viz., X1 and X10. In the X1 position full gain is obtained, and the input
signal is fed directly to the vertical amplifier. However, on the X10
position, the signal amplitude is reduced to 1/10 of its first value.

Let us consider the R and C values that are involved. If the input
resistance of the divider in Fig. 3–15 is to be 1 megohm, a 10-to-1 voltage
division requires that R1 equal 0.9 megohm, and R2 equal 0.1 megohm.
Similarly, since R1C1 must equal R2C2, it follows that C2 must equal 9
times C1. It is evident from previous discussion that the total value of
capacitance shunting R1 is equal to the stray capacitance that is present,
plus C1. Also, the total value of capacitance in shunt with R2 is equal
to the stray capacitance that is present, plus C2.

In conclusion, a compensated input attenuator is necessarily of the
step-switch type, in order to establish fixed values of stray capacitance in
the attenuator network. A typical service scope employs more than one
compensated voltage divider, in order to provide a wide range of attenuation. For example, the majority of service scopes utilize a step attenuator
with ×1, ×10, ×100, and ×1000 positions.

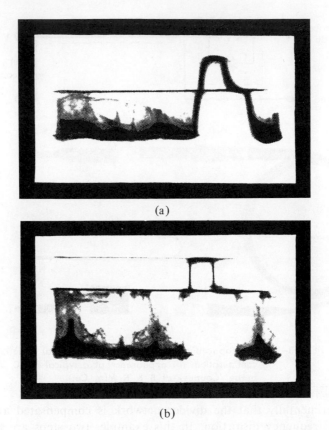

(a)

(b)

Fig. 3–17   Effect of low-capacitance probe adjustment on
waveform: (a) sync pulse distorted by subnormal
probe capacitance; (b) true waveform produced
with correct probe capacitance

Although a compensated step attenuator eliminates frequency distortion in the attenuator network, we are still concerned with the high value
of input capacitance when a shielded input cable is used. Accordingly, a
low capacitance probe such as depicted in Fig. 3–16 is always used in
TV servicing procedures. This is essentially a compensated external step
for the vertical input attenuator. As such, the probe introduces signal
attenuation (customarily 10-to-1), but also reduces the input capacitance
of the cable to approximately 1/10 of its initial value. In other words, a
low-capacitance probe entails a trade-off in which a signal reduction of
90% is accepted in return for an increase of 10 times in the scope's
input impedance. The signal loss in the probe is recovered in the high-gain
vertical amplifier that is utilized in modern service scopes.

As a practical observation, it should be noted that distortionless reproduction of complex waveforms requires that trimmer capacitors such as C1 and C2 in Fig. 3–15 must be correctly adjusted, and also that the trimmer capacitor in the low-capacitance probe, such as C in Fig. 3–16, be correctly adjusted. Once precise adjustments are made, no further attention is required unless a component is replaced. In case the input step attenuator is properly compensated, but the probe trimmer capacitor is set slightly too low, a sync pulse from a TV circuit will be distorted as illustrated in Fig. 3–17. To anticipate subsequent discussion, attenuator and probe adjustments are made to best advantage with a square-wave generator signal.

3.6 DEMODULATOR PROBES AND APPLICATIONS

Another type of oscilloscope probe that finds considerable application in servicing procedures is called a demodulator probe. A typical configuration is shown in Fig. 3–18. This probe is sometimes called a *traveling*

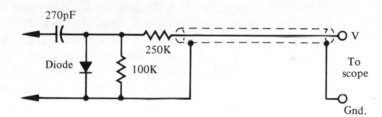

Frequency Response Chart

RF carrier range	500kHz to 200MHz
Modulated - signal range	30 to 5,000 Hz

Input resistance (approx.)

At 500kHz	25k ohms
1MHz	23k ohms
5MHz	21k ohms
10MHz	18k ohms
50MHz	10k ohms
100MHz	5k ohms
150MHz	4.5k ohms
200MHz	2.5k ohms

Maximum input:

AC voltage	20 RMS volts
	28 peak volts

Fig. 3–18 Configuration and specifications for a demodulator probe

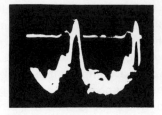

Fig. 3–19   Demodulator probe distorts TV station signal when
signal-tracing the IF amplifier

detector. That is, it can be used to trace a high-frequency signal step-by-step through a receiver section, such as an IF amplifier. A demodulator probe effectively extends the bandwidth of an oscilloscope to VHF frequencies, because the probe processes the input signal and develops the comparatively low-frequency modulation envelope for energizing the vertical amplifier. Although a demodulator probe distorts a TV station signal appreciably, as illustrated in Fig. 3–19, it is very useful to test an IF amplifier for the point where signal blocking occurs.

A demodulator probe is also used in certain visual alignment procedures as explained in greater detail subsequently. In this application, the probe demodulates the sweep-frequency signal for display of the envelope as a frequency-response curve on the scope screen. We will also find that a demodulator probe is useful in frequency-calibration procedures, as discussed in the signal-generator chapter. Another useful application for the probe is in checking the output from a sweep-frequency generator, as described subsequently.

3.7 PRINCIPLES OF TRIGGERED
TIME BASES

Triggered time bases were mentioned previously, and a typical application was illustrated wherein the color burst is picked out from the complete color signal for expansion on the CRT screen. We will find service-type scopes that employ a triggered time base for all waveform displays. These instruments are called triggered-sweep scopes. We will also find service-type scopes that provide a choice of triggered or recurrent time-base operation. This type of instrument is technically termed a synchroscope. For example, the sweep selector depicted in Fig. 3–3 is basically a synchroscope control. We observe that three "Sawtooth" or recurrent positions are provided. In addition, three triggered positions are provided. Note that the vernier or "Sweep Frequency" control is inoperative in the triggered mode of operation. That is, the time base operates precisely at

100 microseconds per centimeter, or 9 microseconds per centimeter, or 1 microsecond per centimeter in the triggered mode. In other words, a triggered time base is precisely calibrated.

Calibrated sweeps enable the operator to measure elapsed time intervals in waveforms, as exemplified in Fig. 3–20. The leading edge of the waveform has been picked out and expanded horizontally for measurement of its rise time. Rise time is defined as the elapsed time from 10% to 90% of peak amplitude on the leading edge. Similarly, the fall time of a waveform is defined as the elapsed time from 90% to 10% of peak amplitude on the trailing edge. This type of measurement is important, for example, in troubleshooting driver circuits in solid-state TV horizontal output sections. Figure 3–21 shows the result of progressively increasing the speed of a triggered time base to make rise-time measurements. We observe that at 0.02 millisecond per centimeter, the leading edge is not expanded sufficiently to measure its rise time. On the other hand, at 0.04 microsecond per centimeter, the leading edge is expanded horizontally to an extent that permits practical rise-time measurement. The rise time in this example is approximately 0.02 microsecond.

A block diagram for a typical service-type triggered-sweep scope is shown in Fig. 3–22. The sectional arrangement is similar to that for the recurrent-sweep block diagram in Fig. 3–1. However, we note that the sweep system includes a trigger section, and that an unblanking amplifier is provided to bring the CRT out of cutoff. The sweep generator remains inactive until a trigger pulse is applied. Thereupon, one cycle of sawtooth voltage is generated, and the electron beam makes one horizontal traverse on the CRT screen. After the traverse is completed, the sweep generator

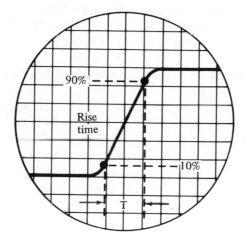

Fig. 3–20   Rise-time measurement on the leading edge of a waveform

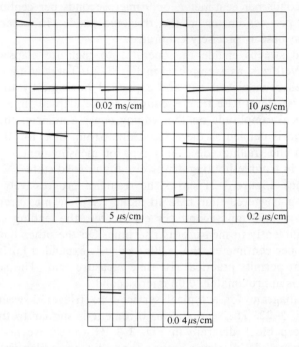

Fig. 3–21 Expansion of the leading edge of a pulse for
rise-time measurement

again remains in a quiescent state until another trigger pulse is applied.
An unblanking amplifier is required because the CRT beam is operated
at very high intensity when displaying waveforms at high writing rates.
That is, the CRT would be damaged if the electron beam were not extin-
guished during the time that it is not being swiftly deflected across the
screen. Thus, the unblanking amplifier switches the beam on only for the
duration of a sawtooth cycle.

Figure 3–23 depicts the configuration for the vertical amplifier in a
triggered-sweep scope. This circuit arrangement is essentially the same as
that employed in a recurrent-sweep scope. In this example, a DC amplifier
is utilized with a bandwidth of 10 MHz. The gain is sufficient to provide
a deflection sensitivity of 10 mV/cm. Push-pull (balanced) amplifier stages
are used throughout, for two basic reasons. First, DC amplifier drift is
thereby minimized because the circuit is essentially a bridge arrangement.
Second, optimum amplitude linearity is provided by push-pull amplification. We observe that the input stage in Fig. 3–23 employs a pair of
FET's. Thereby, a very high input impedance is obtained. TR21 and
TR22 operate as a paraphase inverter, changing the single-ended input
signal into a double-ended (push-pull) output signal.

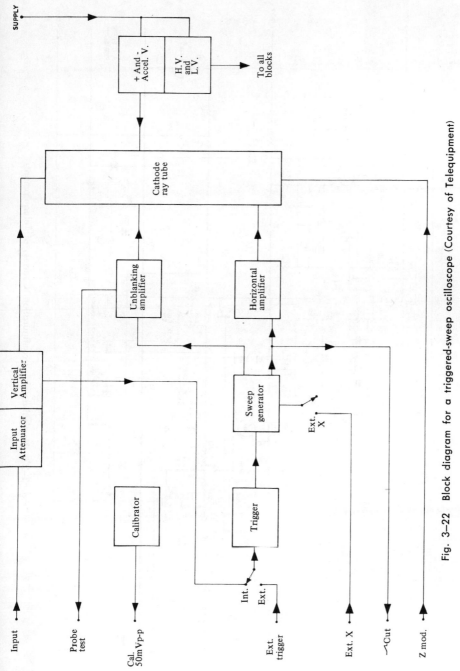

Fig. 3–22 Block diagram for a triggered-sweep oscilloscope (Courtesy of Telequipment)

85

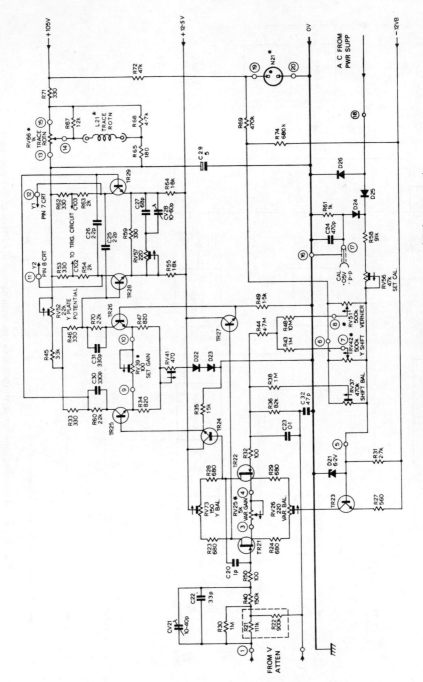

Fig. 3–23 Vertical amplifier configuration (Courtesy of Telequipment)

Four vertical amplifier stages are provided in Fig. 3–23. With the exception of the input stage, junction transistors are employed. A 10-MHz bandwidth is obtained by utilizing comparatively low values of collector load resistance, and by high-frequency feedback via C27 and CV28. TR24 and TR27 are operated as emitter followers, to match the high-impedance input stage to the medium-impedance driver stage. Voltage regulation is provided by TR23 and D21, so that line-voltage fluctuation has negligible effect on amplifier action. We observe that a compensated RC network is used in series with the input lead to TR21. This is similar to a low-capacitance probe, but its purpose is to provide a high-impedance driving source for the FET, thereby protecting it against accidental overload damage.

Diodes D22 and D23 in Fig. 3–23 contribute to temperature stabilization. D24, D25, and D26 produce a 60-Hz square-wave output for calibration of the vertical amplifier. A sample of the signal is taken from the collector loads of TR28 and TR29, for processing by the time-base trigger section. RV66 is a trace-rotation control, which serves as a leveling adjustment for the horizontal trace in the CRT. That is, current flow through L21 produces a magnetic field that produces more or less rotation of the screen pattern clockwise or counterclockwise. Neon bulb N21 serves as a pilot light and as a voltage regulator for the shift (centering) controls.

From the vertical amplifier output stage, the vertical-deflection plates of the CRT are driven in push-pull. This balanced drive provides optimum focus and minimum waveform distortion. The vertical-signal sample proceeds to the trigger circuit where it is mixed with the DC bias voltage for a one-shot multivibrator. In turn, the multivibrator is fired and a sharp pulse is generated at a chosen point along the signal waveform. This pulse is then fed to the triggered time-base section which is a one-shot sawtooth generator. In turn, a single cycle of sawtooth deflection voltage is generated. At the same time, the time-base circuit applies an unblanking pulse to the CRT grid. An electronic switch is included. This is called a lockout or hold-off circuit. It serves to prevent the time base from triggering again until the sawtooth has fallen to zero. Thereby, a small section of the vertical signal can be displayed on the CRT screen without overlapping due to false triggering or multiple triggering.

In conclusion, it is of interest to note the ten key waveforms that are checked in color-TV receivers. These are the video detector output, video output, bandpass amplifier output, burst amplifier output, chroma demodulator outputs, sync separator output, AGC keyer input and output, chroma amplifier outputs, chroma matrix output, and power supply waveforms. A triggered-sweep oscilloscope will provide maximum waveform information in these tests.

QUESTIONS AND PROBLEMS

True-False

1. The sensitivity of a CRT is about 30 volts per inch.
2. The gain of the horizontal amplifier in a typical oscilloscope is the same as the vertical gain.
3. The high voltage for operation of a CRT must furnish a large current.
4. The time base of an oscilloscope is developed by a sawtooth waveform.
5. Most technicians prefer to use retrace blanking in sweep-alignment tests.
6. An oscilloscope cannot be used to measure DC voltages.
7. The peak-to-peak voltage measured by an oscilloscope is true regardless of the shape of the waveform.
8. Stray field pickup becomes an annoying problem when open test leads are used to test television circuits.
9. The use of a shielded cable causes no harmful effects on the circuit under test.
10. A typical service scope provides several attenuator positions.
11. A 10-to-1 probe reduces the signal to 1/10 and the capacitance effect to 1/10.
12. A demodulator probe is a low-capacitance probe that does not reduce signal amplitude.
13. A demodulator probe is useful in frequency-calibration procedures.
14. A synchroscope does not provide a triggered sweep.
15. A triggered time base should be precisely calibrated.
16. The fall time of a square wave is the time elapsed between 90% and 10% of peak amplitude on the trailing edge.
17. The sweep of an oscilloscope is the result of the application of an internally generated sawtooth to the horizontal plates.
18. The circuit in Fig. 3–22 uses a blanking amplifier to prevent the appearance of the retrace on the CRT.
19. The hold-off circuit in Fig. 3–23 prevents the time base from triggering again until the sawtooth has fallen to zero.
20. A small portion of a waveform can be viewed on the CRT of a triggered oscilloscope.

Multiple-Choice

1. A voltage gain of _____ is typical for the horizontal amplifier.
   (a) 1
   (b) 30
   (c) 1500
2. The brightness of the trace on the CRT in an oscilloscope is controlled by the _____ control.
   (a) intensity
   (b) focus
   (c) brightness

3. The purpose of the sync control is to
    (a) set the intensity level.
    (b) set the focus.
    (c) lock the signal.
4. The length of the sweep on the CRT is controlled by
    (a) sync control.
    (b) vertical gain.
    (c) horizontal gain.
5. If the retrace is visible on the CRT, the trouble may be
    (a) intensity too high.
    (b) defective component in the retrace blanking circuit.
    (c) loss of signal sync.
6. An oscilloscope can be calibrated to measure _____ voltages.
    (a) AC
    (b) DC
    (c) AC and DC
7. To prevent loading of a circuit under test, the input impedance of the oscilloscope must
    (a) be low.
    (b) be high.
    (c) appear capacitive.
8. The best method for testing waveforms of a television is to use an oscilloscope with
    (a) open test leads.
    (b) a shielded test cable.
    (c) a low-capacitance probe.
9. The low-capacitance probe
    (a) reduces the input signal to the vertical amplifier.
    (b) increases distortion.
    (c) causes loading.
10. A typical low-capacitance probe has an attenuation ratio of
    (a) 2-to-1.
    (b) 5-to-1.
    (c) 10-to-1.
11. Attenuator and probe adjustments are made to best advantage with a $(-)$ signal
    (a) sine wave.
    (b) pulse.
    (c) square wave.
12. A demodulator probe is used to
    (a) extend the bandwidth of the oscilloscope.
    (b) increase the gain of an oscilloscope.
    (c) prevent distortion of the signal.
13. A useful application of the demodulator probe is
    (a) checking the output of a sweep-frequency generator.
    (b) observing a very low amplitude audio signal.
    (c) observing a square wave.
14. An oscilloscope that provides a choice of triggered sweep or recurrent time-base operation is called a
    (a) triggered-sweep scope.

(b) synchroscope.
(c) vectorscope.

15. The rise time of a square wave is the time elapsed between _____ of peak amplitude.
    (a) 0 and 100%
    (b) 10 and 90%
    (c) 50 and 100%

General

1. What is a typical voltage gain of the horizontal amplifier of an oscilloscope?
2. List the following controls and describe their functions: focus, intensity, vertical centering, horizontal centering, and sweep selector.
3. How should the horizontal control be adjusted?
4. Why must the horizontal time base be linear?
5. What is the purpose of retrace blanking?
6. What is the purpose of the "Int.-Ext." switch?
7. What values of voltage is the oscilloscope usually used to measure?
8. What is the difference between the DC and the AC calibration of an oscilloscope?
9. What is the purpose of the low-capacitance probe?
10. Why should you use shielded cable for the input leads to an oscilloscope?
11. How is the frequency-compensating capacitor adjusted for optimum frequency response of the vertical amplifier?
12. What is the usual division of the input voltage from the probe input to the vertical input?
13. What is the purpose of the demodulator probe?
14. What is the purpose of the calibrated time base?
15. How are the rise and fall times of a square wave defined?
16. Why is the CRT in Fig. 3–23 driven in push-pull?

4

AM Signal
Generators

4.1 SURVEY OF AM SIGNAL GENERATORS

Among the various types of signal generators utilized in servicing procedures, the amplitude-modulated (AM) RF signal generator is the most basic. (See Fig. 4–1.) The AM signal generator has numerous applications, among which are alignment of AM and FM radio receivers, alignment of TV receivers, signal-injection tests in troubleshooting procedures, stage-gain measurements, resonant-frequency measurements, frequency calibration of auxiliary equipment, measurement of tuned-circuit Q values, and so on. We will find that an AM signal generator is better adapted to some kinds of applications than to others. For example, an FM radio receiver can be aligned more conveniently and faster with a specialized FM signal generator than with an AM generator. Again, a TV receiver can be aligned to better advantage with specialized TV signal generators than with an AM signal generator.

Frequency ranges, amplitude-modulating facilities, calibration accuracy, output-voltage range, and waveform purity may differ considerably from one class of AM signal generator to another. (See Fig. 4–2.) Thus, one generator might cover a frequency span from 85 kHz to 40 MHz, whereas another generator might cover a span from 100 kHz to 20 MHz. Still another generator might cover a frequency span from 100 kHz to 60 MHz. Still another generator might cover a frequency span from 160 kHz to 110 MHz, with a harmonic scale calibration from 110 MHz to 220 MHz. We will find that applications are generally more straightforward and reliable when frequency tests are made with the fundamental output from a

Fig. 4–1   Appearance of an AM signal generator (Courtesy of EICO)

generator. Harmonic tests are regarded only as an expedient when a suitable generator is unavailable. These considerations are examined in greater detail subsequently.

Most AM signal generators have built-in 400-Hz amplitude-modulating facilities, and some instruments also provide 1-kHz modulation. However, most service-type generators are equipped with built-in amplitude modulators with "External Modulation" terminals, whereby a supplementary audio oscillator can be used to modulate the RF signal at any audio frequency. Most AM signal generators also provide output terminals for the internal 400-Hz or 1-kHz modulating voltage. In turn, this audio signal can be used for injection tests in audio amplifiers, and in similar applications. A service-type AM signal generator may provide only a fixed percentage of modulation (usually 30%), or some instruments may have a percentage-modulation control. As explained in greater detail subsequently, the modulation percentage is limited to approximately 30% in instruments that employ modulation of the RF oscillator. Examples of amplitude modulation are shown in Fig. 4–3.

Output voltage ranges have a maximum value of at least 50,000 microvolts (0.05 volts) in most service-type AM generators. Some instruments provide a maximum output of 1 volt. However, the maximum output is sometimes available only on low-frequency ranges. For example, a generator might provide a maximum output of 1 volt at 100 kHz, with a maximum output of 0.1 volt at 50 MHz. Harmonically calibrated frequency ranges can provide only a fraction of the maximum output available on fundamental ranges. Attenuators provided on most service-type generators can attenuate the output signal down to several microvolts, particularly on the lower-frequency ranges. On the higher-frequency ranges, the minimum output is dominated by RF leakage from the instrument case and radiation from the power cord.

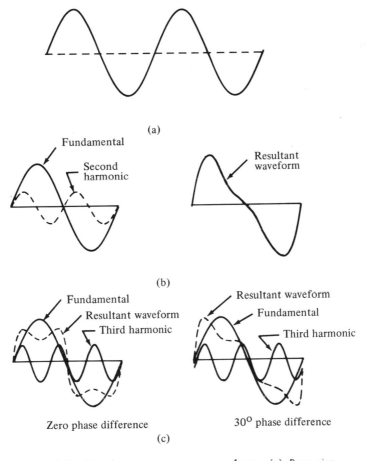

Fig. 4–2   Signal-generator output waveforms: (a) Pure sine waveform; (b) Second-harmonic distortion; (b) Examples of third-harmonic distortion

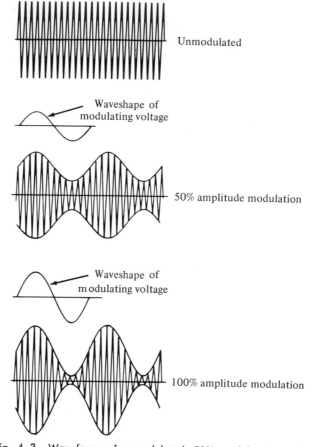

Fig. 4–3   Waveforms of unmodulated, 50% modulated, and 100% modulated output signals

4.2   AM SIGNAL GENERATOR FEATURES

All signal generators contain an oscillator, calibrated tuning capacitor, and attenuator, as depicted in Fig. 4–4. An output meter is provided in the more elaborate instruments. A calibrated attenuator is employed when an output meter is utilized. In turn, the number of microvolts in the output signal is indicated. The output meter is sometimes called a carrier-level meter; if the pointer is set to the reference carrier-level mark on the scale, a signal amplitude of 1 volt might be applied to the attenuator, for example. Since the attenuator is decimally calibrated, the output amplitude from the attenuator can be adjusted for a level of 10, 20, 50, 100, 1000, or 10,000 microvolts, for example. However, utility-type RF generators usually have uncalibrated attenuators and dispense with an output meter. In the example of Fig. 4–4, the oscillator section is double-shielded, and a third shield is provided around the attenuator. Extensive shielding makes it possible for the operator to work with very low output levels, such as 1 or 2 microvolts.

Figure 4–5 shows the sectional layout of a comparatively elaborate service-type AM generator. Note that the audio modulating voltage is applied to a buffer stage instead of the RF oscillator stage. This method of modulation provides two operating advantages. First, the modulation level can be set for 100% modulation, if desired. However, if the oscillator were directly modulated, its operation would become erratic at higher percentages of modulation. Second, if an oscillator is directly modulated, it develops appreciable *incidental FM* when operated at appreciable percentages of modulation. In other words, the average frequency varies back and forth between modulation peaks. This spurious incidental-FM output is undesirable in the majority of applications. Therefore, it is good practice to modulate the buffer stage instead of the oscillator stage.

As its name indicates, a buffer stage serves to isolate the oscillator from subsequent circuit sections. The buffer stage seldom provides appreciable voltage gain, although it may provide substantial power gain. The output from the buffer may be fed into the attenuator network, or, if a comparatively high-level output is desired, a broad-band amplifier may be placed between the buffer and the attenuator. In the example of Fig. 4–5, a three-position ladder attenuator is employed. Switch S2 can be set to the approximate level that is desired, and the fine attenuator control is adjusted to "fill in" between the steps of the ladder or coarse attenuator. The fine attenuator is essentially an amplifier gain control. Note that

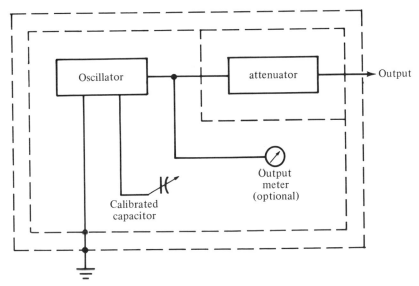

Fig. 4–4   Plan of the basic RF generator

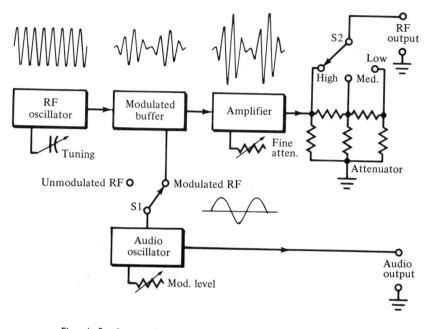

Fig. 4–5   Sectional layout of a comparatively elaborate AM generator

the audio oscillator output is available at the audio output terminals. Switch S1 has two positions, whereby the operator can choose either modulated or unmodulated RF output.

Most service-type AM generators employ the basic Hartley oscillator configuration depicted in Fig. 4–6(a). This is a positive-feedback arrangement with the transistor operating in the common-base mode. A shunt-fed configuration is often employed, as depicted in Fig. 4–6(b). R_B, R_C, and R_F are bias resistors. C2 is a blocking capacitor, and C_E is an emitter bypass capacitor. C_C operates as a coupling capacitor from T1 to the base of Q. In this example, the transistor operates in the common-emitter mode. Next, the series-fed configuration shown in Fig. 4–6(c) is utilized in other signal generators. R_B and R_F are bias resistors. C_E is an emitter bypass capacitor, and C2 provides a low-impedance AC path around the power source. C_C operates as a coupling capacitor from T1 to the base of Q, which operates in the common-emitter mode.

Band switching as exemplified in Fig. 4–7 is typical. This arrangement has five bands with a frequency span from 100 kHz to 20 MHz. The coil inductances range from 10 mH to 1μH. A tuning capacitor with a maximum capacitance of 250 pF is utilized on each range. Approximately 4 volts of RF output into a 500-ohm load is obtained. Note that resistors R5 through R9 are placed in the feedback circuit to equalize the output on

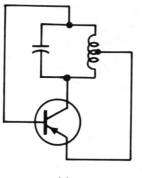

(a)

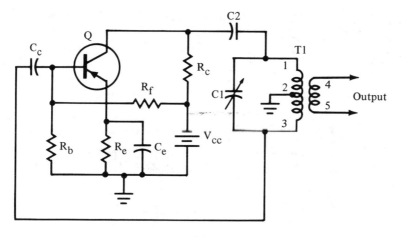

(b)

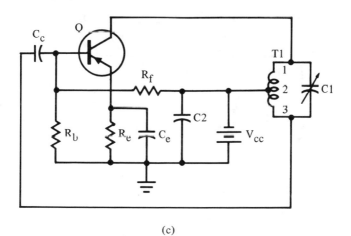

(c)

Fig. 4–6 Hartley oscillator circuits: (a) basic configuration;
(b) shunt-fed circuit; (c) series-fed circuit

each band and also to linearize the emitter circuit of the transistor for generation of a good sine waveform. R8 and R9 have lower values than the other feedback resistors because the Hartley configuration has effectively looser coupling between the tapped portions of the tank coil at higher operating frequencies.

Most service-type AM generators employ a directly modulated oscillator, as depicted in Fig. 4–8. The 400-Hz audio oscillator Q2 in this example generates a sine-wave voltage which is superimposed on the DC supply voltage to Q1, via T1, thereby providing amplitude modulation. A fixed modulation depth of approximately 30% is obtained on all six bands (only two RF bands are shown in the diagram). Note that the audio oscillator employs circuitry characteristic of both tickler feedback and Colpitts configurations. In turn, the Colpitts network provides a good sine waveform, and the tickler branch couples additional feedback voltage to the transistor to obtain increased output. We observe that a simple RF attenuator is utilized; it consists of a 400-ohm potentiometer in the emitter

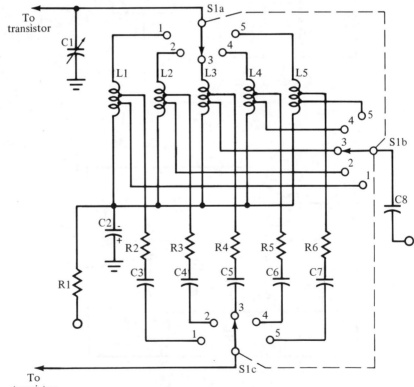

Fig. 4–7 Oscillator band-switching arrangement

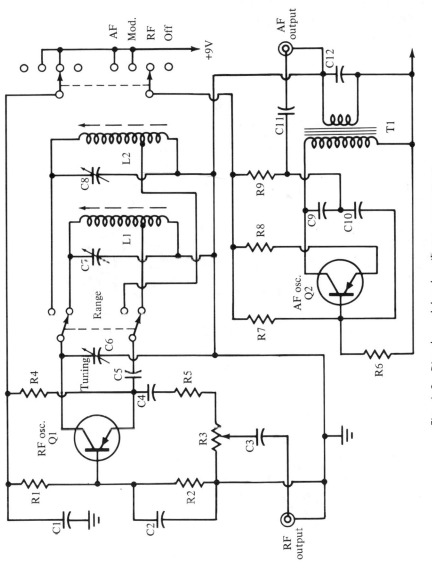

Fig. 4–8 Directly modulated oscillator arrangement

circuit of Q1. Maximum RF output is approximately 50,000 microvolts, and the minimum output is low enough so that all-wave broadcast receivers can be aligned satisfactorily. However, communication-receiver alignment requires more elaborate generator configurations, such as depicted in Fig. 4–5.

4.3   SERVICE-SHOP APPLICATIONS

Alignment of tuned circuits in AM broadcast receivers is accomplished to best advantage with the test setup shown in Fig. 4–9. To provide normal

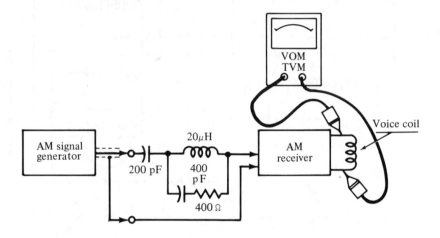

Fig. 4–9   Test setup for alignment of an AM broadcast receiver

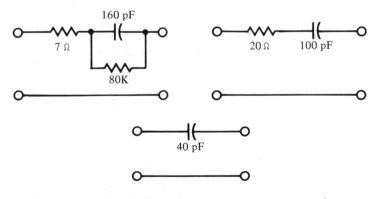

Fig. 4–10   Simple dummy antenna arrangements used in auto-radio alignment procedures

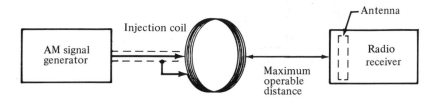

(a)

Shape a loop of several turns of wire and connect the generator across the loop. Adjust the output level of the generator to radiate just enough energy to be received.				
Generator frequency	Dial setting	Indication	Adjust	Remarks
1 455KHz unmod.	Tuning cap.	DC voltage across volume control	IF trans.	Adjust for maximum
2 1020KHz	··	··	Osc. trimmer	Adjust for maximum
3 600KHz	Tune to signal	··	Osc. slug	Rock tuning gang. Repeat 2 and 3
4 1400KHz	··	··	Trimmer	Adjust for maximum

(b)

Fig. 4–11 Alignment setup for receiver with built-in loop antenna: (a) injection coil couples test signal into receiver; (b) typical alignment instructions for a simple receiver (Courtesy of Howard W. Sams & Co., Inc.)

loading of the receiver antenna input section, a standard dummy antenna is utilized. This consists of a series-parallel RCL network that stimulates the average radio antenna system. A VOM or TVM is connected across the speaker voice-coil terminals, to serve as an output meter. Generally, very low RF output is used from the signal generator so that the AVC system of the receiver is practically "wide open." This procedure ensures that the tuned circuits are in optimum alignment for weak-signal reception. Modulated RF output must be used in this test setup. Details of alignment procedure are usually available in the receiver service data.

Simpler dummy antenna arrangements are commonly used in auto-radio alignment procedures, as exemplified in Fig. 4–10. The recommended configuration is usually specified in the receiver service data.

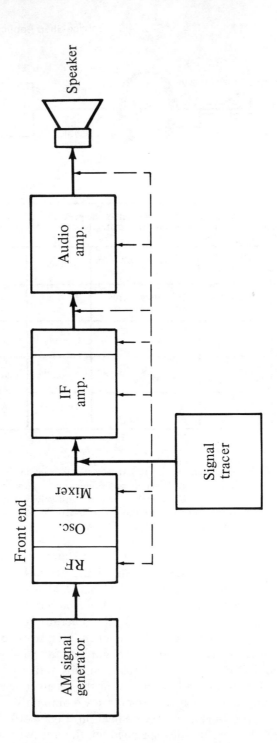

Fig. 4–12 Signal-tracing test setup

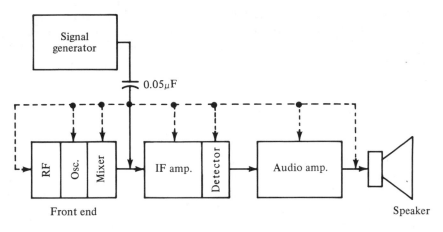

Fig. 4–13   Signal-substitution test arrangement

When a receiver operates from a built-in loop antenna, the signal application setup in Fig. 4–11(a) is generally utilized. The generator output signal is applied to a signal-injection coil which consists of several turns of wire. This coil is then placed at a distance from the receiver under test such that a readable indication is produced on the output meter. Typical alignment instructions for a simple receiver are given in Fig. 4–11(b).

Alignment is always accomplished last, in case of a defective receiver. The only exception is when the technician knows that receiver malfunction has been caused by tampering with the alignment adjustments. There are two general methods of using signal generators for troubleshooting defective receivers. These methods are called signal-tracing and signal-substitution procedures. A signal-tracing test setup is shown in Fig. 4–12. The AM signal generator applies a modulated RF signal to the receiver input terminals, and the receiver is tuned to the same frequency as the generator. A signal tracer is then applied successively at the RF, mixer, IF, detector, and audio stages to check for presence or absence of signal and its comparative strength. Thereby, a defective stage can be quickly and easily located. Signal tracers are described in greater detail subsequently.

Signal-substitution procedure is analogous to signal-tracing procedure, except that the various stages are checked by signal injection instead of probing with a signal tracer. Hence, a signal tracer is not required when this method is used. Figure 4–13 shows the setup employed for signal-substitution tests. The first test is ordinarily made at the speaker, using audio-frequency output from the generator. If a tone is heard from the speaker, the injection point is moved back through the audio amplifier. In case the audio tone continues to be heard, the injection point is moved back through the IF section, using modulated IF output from the generator. In case the audio tone still continues to be heard, the injection point is moved back through the front end, using modulated RF output. Note that the oscillator is tested by using an unmodulated RF output of suitable frequency from the generator. Thus, the generator can substitute for a "dead" local oscillator. Of course, in an oscillator check, the receiver must be energized from an antenna or from another signal generator.

Since each section of a receiver normally provides amplification, maximum generator output is used to drive the audio output section, whereas minimum generator output is applied to the input of the RF amplifier. An oscillator check usually requires maximum generator output. As shown in Fig. 4–13, an output blocking capacitor of approximately 0.05μF should be connected in series with the "hot" lead from the generator. This blocking capacitor ensures that DC bias voltages in various receiver sections will not be drained off through the generator output cable and attenuator. Note that although preliminary tests can be made by utilizing the speaker as an indicator, stage-gain measurement requires that a VOM or TVM be connected at the audio-amplifier output to serve as an output meter. In turn, if a given signal from the generator is applied at the output of a stage and then at the input of the stage, the voltage gain of the stage is measured by the difference in the two readings on the meter.

Stage gains in RF and IF stages can be measured accurately if the AVC system of the receiver is disabled. That is, if the AVC system is operative, the gain of a stage will decrease when the signal-injection point is changed from the output to the input point of the stage. Most receivers operate at maximum gain when the AVC system biases germanium transistors at about 0.25 volt, or silicon transistors at about 0.65 volts. The AVC section of a receiver is customarily "clamped" at a fixed operating point by shunting a low-impedance bias source across the AVC supply terminals. This fixed bias can be obtained from a commercial bias box, or can be improvised from a flashlight battery and a low resistance potentiometer (25 ohms, for example). Stage-gain tests are most useful when troubleshooting a "weak receiver" symptom. In case of doubt concerning acceptable gain values, it is sometimes possible to make comparative tests in another normally operating receiver of the same type.

Figure 4–14 shows how to measure the Q value of a tuned stage. The generator signal is applied to the base of the transistor preceding the tuned circuits to be tested. The primary and secondary of the following transformer are shunted with 200-ohm resistors, to "kill" their resonant response. A TVM with an RF probe is connected at the collector of the transistor following the tuned circuits to be tested. The generator output

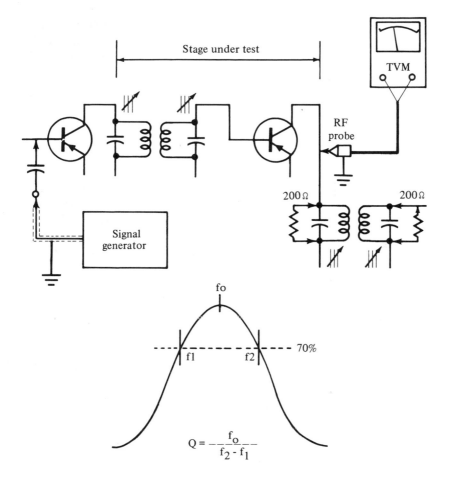

Fig. 4–14   Measurement of tuned-stage Q value

is adjusted to a level that does not overload the stage when operating at its resonant frequency f_0. Then the generator is tuned to a lower frequency f_1, at which the TVM indicates 0.707 of the voltage that is indicated at f_0. Finally, the generator is tuned to a higher frequency f_2, at which the TVM indicates 0.707 of the voltage that it indicated at f_0. To find the Q value of the stage, we calculate the difference between f_2 and f_1, then divide this difference into f_0. For example, if f_0 measures 465 kHz, f_1 measures 460 kHz, and f_2 measures 470 kHz, we divide 10 into 465, and find that the Q value is 46.5.

An AM signal generator is often used to test and align FM radio-receiver circuitry. Except for the higher operating frequencies employed by the FM circuits, the general procedures are much the same as in AM receivers. However, the FM detector section is more elaborate than an AM detector section, and an FM detector includes tuned circuits that require individual attention. The ratio-detector type of FM detector is most widely used, as depicted in Fig. 4–15. In an FM broadcast receiver, the detector circuits operate at a center frequency of 10.7 MHz. Figure 4–16 shows the normal frequency-response curve for a radio detector. We observe that the detector has zero output at the center frequency of 10.7 MHz, with one peak response at 10.6 MHz, and another peak response at 10.8 MHz. Peak output voltages of ±12 volts are typical.

A suitable test setup for checking a ratio detector is shown in Fig. 4–15. The signal generator is operated on unmodulated RF output at 10.7 MHz, and the test signal is capacitively coupled to the base of the transistor preceding the ratio detector. Volume control R3 should be set to maximum. A TVM is connected across the de-emphasis capacitor C7. Maximum output level from the generator is generally utilized in ratio-detector tests. If the tuned circuits are in correct alignment, the TVM will indicate zero output at the test frequency of 10.7 MHz. If a positive or negative DC voltage is indicated, we adjust the secondary tuning

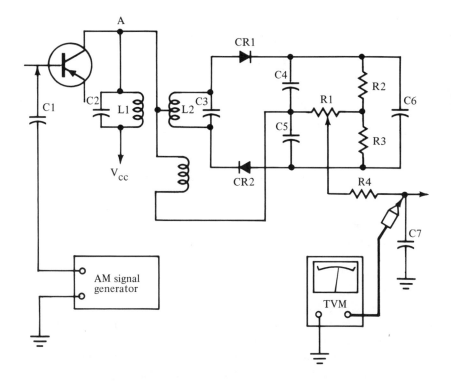

Fig. 4–15   Test setup for checking a ratio detector

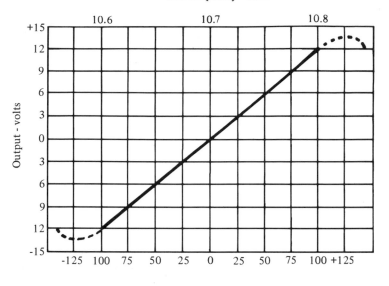

Fig. 4–16   Normal frequency-response curve for a ratio detector

capacitor C2 for zero output. Next, the signal generator is set to 10.6 MHz and the TVM reading noted. This reading can be maximized by adjustment of the primary tuning capacitor C1. Then the signal generator is set to 10.8 MHz and the TVM reading noted again. Peak readings are normally the same at both 10.6 and 10.8 MHz.

In case trimmer capacitor C1 (Fig. 4–15) requires appreciable adjustment to maximize the TVM reading at 10.6 MHz, it is good practice to recheck the adjustment of C2. That is, we apply a 10.7-MHz signal as before. If zero output is not indicated, the adjustment of C2 is reset as required. Thereby, balanced output and minimum distortion are obtained. Inability to align the tuned circuits for normal frequency response is usually the result of a defective capacitor. However, if no capacitor faults are found, it is advisable to check diodes X1 and X2 for front-to-back ratios. Optimum operation is ensured by using a pair of matched diodes— that is, diodes that have closely similar front and back resistance values. Finally, we should note that although a ratio-detector circuit is in correct alignment, noisy and distorted operation can result from an open stabilizing capacitor (C6 in Fig. 4–15).

Ratio detectors in TV receivers are basically the same as in FM broadcast receivers, except that the tuned circuits operate at 4.5 MHz.

Also, the deviation of a TV sound signal is ±25 kHz, instead of ±75 kHz, as in FM broadcast transmission. Accordingly, the ratio detector in a TV receiver normally has zero output at a test frequency of 4.5 MHz, and peak outputs at 4.25 and 4.75 MHz. In practice, we will often find that the peak responses occur at the ends of somewhat greater frequency separation. The essential point is that the peak responses must not occur at a frequency higher than 4.25 MHz, nor at a frequency lower than 4.75 MHz. As noted previously, each peak should produce the same DC voltage indication on the TVM. In summary, the ratio detector should provide a full 50-kHz bandwidth, with equal positive and negative peak outputs.

4.4 AM GENERATOR CALIBRATION

From time to time, the calibration of a signal generator should be checked against accurate frequency standards. Then, if it is found that the scale indication has drifted, a tabulation can be made or a correction graph can be drawn, so that the instrument can be set to a precise operating frequency for tests in critical circuits. Broadcast station signals are generally the most convenient source of accurate frequency standards. A suitable test setup for checking generator calibration is depicted in Fig. 4–17. If an all-wave radio receiver is used, spot checks of generator calibration can usually be made on all bands. Note that an FM receiver or a TV receiver can be used to check the generator calibration against additional carrier frequencies. A small capacitor is used to couple the

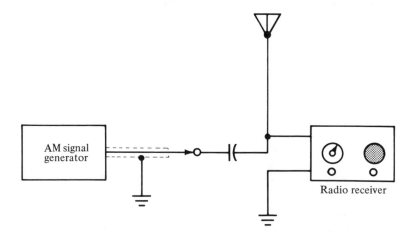

Fig. 4–17   Generator calibration arrangement

Station	Location		Carrier freq. MHz	Radiated power KW	Antennas, modulation
WWV	Fort Collins Colorado				Radiation from all antennas is ommidirectional, from vertical halfwave dipoles except the 2.5MHz and the 5MHz antennas at WWVH, which are quarter-wave dipoles.
	40°45 55 .2N	105°02 31 .3W	2.5	2.5	
	40°40 42 .1N	105°02 25 .9W	5	10	
	40°40 47 .8N	105°02 25 .1W	10	10	
	40°40 45 .0N	105°02 28 .5W	15	10	
	40°40 53 .1N	105°02 28 .5W	20	2.5	
	40°40 50 .5N	105°02 26 .6W	25	2.5	
WWVH	Maui, Hawaii				All modulation is double sideband
	Coordinates				
	Lat.	20°46 02 N	5	2	
	Lon.	156°27 42 W	10	2	
			15	2	

Station	Frequency	Accuracy as transmitted	Remarks
WWV and WWVH	440 Hertz	5 parts in 10^{10} or better	Changes in the transmitting medium (Doppler effect, etc.) may produce changes in the audio tone.
WWV and WWVH	600 Hertz	5 parts in 10^{10} or better	The 440- and 600 Hertz tones are broadcast alternately both from WWV and WWVH(see schedule of broadcasts)

Fig. 4–18 Standard frequency signals transmitted by WWV

generator signal into the antenna lead. Thereby, the generator signal is mixed with the station signal to which the receiver is tuned.

As the signal-generator frequency is tuned through the carrier of the incoming broadcast-station signal, a heterodyne beat or "squeal" is heard from the speaker in the receiver. When the generator is tuned for zero beat against the station signal, the generator is then operating at exactly the same frequency as the broadcast station. Since station carrier frequencies are known and maintained to a very high accuracy, the foregoing test provides a precise check of scale calibration. Note that the standard frequency transmissions of WWV (Fig. 4–18) can be tuned in almost anywhere in the country with a short-wave receiver. Spot frequency checks of generator calibration can be made at 2.5, 5, 10, 15, 20, and 25 MHz. These WWV signals have higher accuracy than any other standards.

QUESTIONS AND PROBLEMS

True-False

1. The amplitude modulated signal generator is the most basic signal generator used in television and radio servicing.
2. An AM generator is the best type of signal generator to use for alignment of an FM receiver.
3. Tests using harmonic signals should be used only when the fundamental is not available.
4. The modulation percentage is limited to 60% for instruments that modulate the RF oscillator.
5. The maximum output from a signal generator is limited by radiation from the instrument.
6. All signal generators contain an oscillator, calibrated tuning capacitor, and attenuator.
7. A calibrated attenuator is employed with an output meter.
8. Extensive shielding of the oscillator and attenuator is necessary to protect the operator from radiation.
9. The buffer stage isolates the oscillator from the modulator.
10. The buffer stage usually furnishes power gain.
11. When aligning a receiver, a TVM should be used as an input meter.
12. Alignment is usually accomplished first on a defective receiver.
13. Signal tracing of a receiver requires an instrument other than a signal generator.
14. When a signal tracer is used with a signal generator to test a receiver, the signal generator is usually coupled to the antenna.
15. Signal substitution requires no signal tracer.
16. A signal tracer must make use of a detector to trace a signal in an IF stage.

17. A blocking capacitor is usually used in or with a generator to prevent the loss of DC voltage from the device or component under test.
18. Stage gain in RF and IF stages cannot be measured accurately unless the AVC system is operating.
19. Fixed bias for RF stages can be obtained from a bias box.
20. The ratio detector in an FM receiver operates at a center frequency of 10.7 MHz.
21. The output of a ratio detector is set to zero at 10.8 MHz.
22. The frequency deviation of the sound section of a TV is ±75 kHz.
23. The peak output of a ratio detector is a DC voltage.
24. A correction chart can be drawn for a signal generator by checking it against accurate frequency standards.
25. A local FM station is the most accurate broadcast station to use in a generator frequency check.

Multiple-Choice

1. Results are generally more reliable and straightforward when the _____ frequency of the RF signal of an AM generator is used.
   (a) fundamental
   (b) subharmonic
   (c) harmonic
2. The level of the output of an AM generator is controlled by
   (a) a gain control.
   (b) an attenuator.
   (c) percentage of modulation.
3. On the high-frequency ranges, the minimum output is dominated by the _____ the instrument and radiation from the _ _____.
   (a) RF leakage from; power cord
   (b) leakage from the power cord; oscillator
   (c) leakage from the output cable; output cable
4. Extensive shielding of the oscillator and attenuator are necessary to prevent
   (a) radiation at low signals.
   (b) radiation burns.
   (c) signal pickup at low levels.
5. If an audio oscillator is modulated directly at a large percentage of modulation it will cause
   (a) pulling.
   (b) signal reduction.
   (c) frequency shifting.
6. The buffer stage seldom furnishes
   (a) power gain.
   (b) voltage gain.
   (c) oscillator isolation.
7. The alignment procedure for each receiver is
   (a) the same.
   (b) given on the chassis.
   (c) found in the receiver service data.

8. In an oscillator check the receiver must be energized
   (a) from the antenna or another signal generator.
   (b) by the same generator.
   (c) by a TVM.

9. The signal tracer method of testing a receiver requires the use of an
   (a) AM generator.
   (b) AM generator and TVM.
   (c) AM generator and signal tracer.

10. DC voltages at test points in the receiver are isolated from the generator by using a
    (a) resistor.
    (b) blocking capacitor.
    (c) loop coupling.

11. Stage gain test can be made in a receiver by the use of a _____ and a _____.
    (a) speaker; signal generator
    (b) signal tracer; signal generator
    (c) TVM; signal generator

12. Stage gains are most useful when troubleshooting a
    (a) lead.
    (b) receiver with static.
    (c) weak receiver.

13. The Q of a tuned stage is given by the formula
    (a) $(f_1 + f_2) - f_0$.
    (b) $f_0/(f_2 - f_1)$.
    (c) $(f_2 - f_1)/f_0$.

14. The ratio detector is adjusted for peak readings at
    (a) 10.7 MHz.
    (b) 10.6 MHz.
    (c) both 10.6 and 10.8 MHz.

15. The frequency deviation of an FM receiver is _____ KHz and that of a TV sound section is _____ KHz.
    (a) 25; 75
    (b) 75; 75
    (c) 75; 25

16. The peak outputs of a ratio detector is a/an _____ voltage.
    (a) AC
    (b) positive DC
    (c) negative or positive DC

17. The frequency of a signal generator can be checked against a broadcast station by coupling the generator into a radio and tuning the generator for a
    (a) zero beat.
    (b) squeal.
    (c) 400 cycle tone.

18. The most accurate broadcast station signal is
    (a) a local FM station.
    (b) WWV.
    (c) a local AM station.

General

1. What are the usual limits of fixed amplitude modulation?
2. What is the usual modulation frequency of most AM signal generators?
3. What is another name for the output meter?
4. What is the purpose of the extensive shielding of an AM signal generator?
5. Why is the buffer stage usually modulated rather than the oscillator?
6. What happens if the modulator is modulated directly?
7. Why is alignment always the function to be completed after a receiver is repaired?
8. How could you find the Q of an amplifier circuit?
9. How can a ratio detector be checked with an AM signal generator?
10. What is a simple test to check the frequency calibration of an AM signal generator?

5

Capacitor
Testers

5.1 CAPACITOR TEST REQUIREMENTS

Since capacitors deteriorate more often than other components in
radio, television, and general electronic circuitry, technicians routinely
make numerous kinds of capacitor tests. The most basic classification
comprises out-of-circuit versus in-circuit capacitor test procedures. Some
in-current tests are very simple. For example, if a technician suspects
that a coupling capacitor is open, he usually "bridges" the doubtful capacitor with a known good capacitor of the same value and rating. That is,
the leads of the good capacitor are touched to the leads or terminals of
the suspected capacitor, to see if normal operation is resumed. Then, if the
suspicion is confirmed, he replaces the open-circuit capacitor. Quick
tests of bypass and decoupling capacitors can also be made by this
"bridging" test when the trouble symptoms point to the possibility of an
open-circuited capacitor.

Many defective capacitors produce trouble symptoms because of
leakage. That is, the dielectric between the electrodes of the capacitor
loses its insulating property, so that the capacitor effectively becomes
shunted by more or less resistance. For example, if the coupling capacitor
C_c in Fig. 5-1 develops leakage resistance, the effect on circuit action is
the same as if a resistor R_L were connected in parallel with the capacitor.
The value of this leakage resistance might be as high as several megohms,
or as low as a "dead" short-circuit. Coupling-capacitor leakage produces
a change in DC operating voltages. That is, if C_c in Fig. 5-1 becomes
sufficiently leaky to produce trouble symptoms, R_L will "bleed" collector
voltage from Q1 to the base of Q2. In turn, a simple test for suspected
coupling-capacitor leakage can be made by measuring the base bias

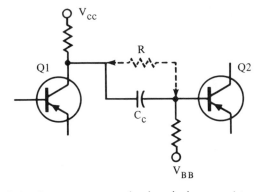

Fig. 5–1   Capacitors may develop leakage resistance, as represented by R_L

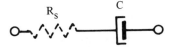

Fig. 5–2   Effective series resistance R_s causes the capacitor to have an excessively high power factor

voltage on Q2, and comparing the measured value with the specified value in the receiver service data.

Coupling capacitors may be of mica, ceramic, paper, or electrolytic dielectric construction. Electrolytic capacitors may cause trouble symptoms by losing a substantial portion of their original capacitance value. This defect can be compared to an open-circuit fault, in that the suspected capacitor can be localized with a "bridging" test. An out-of-circuit test is made by measuring the value of the capacitor on a capacitance bridge. Most service-type capacitor testers are designed with a capacitance bridge and a leakage indicator. A power-factor measuring function is generally also provided for checking the impedance of electrolytic capacitors. As depicted in Fig. 5–2, an electrolytic capacitor can cause trouble symptoms by developing effective series resistance. In turn, the impedance increases. This defect shows up on a capacitor tester as an excessively high power factor, since the power factor is equal to the ratio of resistance to reactance. The power factor of an ideal capacitor is zero.

5.2   FUNCTIONAL SECTIONS OF A CAPACITOR TESTER

Most service-type capacitor checkers provide capacitance-measuring and power-factor measuring functions, leakage indication, and a resistance-bridge function. For example, the instrument illustrated in Fig. 5–3 measures capacitance values from 10 pF to 5000 µF in four ranges, power-factor values from zero to 80%, and resistance values from 0.5 ohm to 500 megohms in four ranges. The leakage-indicator function includes a DC test-voltage control that is adjustable from zero to 500 volts. In addition, the instrument in this example provides a comparator function, which is a simple bridge arrangement for special types of tests. For example, the turns ratio of a power transformer or output transformer can be quickly measured with the comparator function.

Although capacitance values can be measured in more than one way, the majority of service-type capacitor checkers employ the basic capacitance bridge depicted in Fig. 5–4. Note that service-type bridges generally operate at 60 Hz, whereas laboratory-type bridges operate at 1 kHz. Thus, 54 volts at 60 Hz is utilized to energize the bridge in this example. R1 is a current-limiting resistor that prevents damage to the transformer in case the test voltage is applied to a short-circuit. A 200-pF standard capacitor is shown in the diagram, which provides a 10- to 5000-pF range on the calibrated potentiometer. Balance indication is provided by an eye tube, operating in a self-rectifying circuit comprising coupling capacitor C_c and resistor R3. That is, the grid of the eye tube draws current on positive peaks of the applied AC voltage, thereby charging C_c and applying signal-developed bias to the tube. Balance is indicated by maximum opening of the eye sector on the fluorescent screen.

It might be supposed that capacitance values less than 10 pF could be measured with the arrangement of Fig. 5–4. However, this is impractical, due to poor indication accuracy of the ratio-arm potentiometer at very low settings. In practice, if capacitors in the range of 10 to 100 pF are to be checked accurately, the capacitor must be connected directly to the

Fig. 5–3   Service-type capacitor checker (Courtesy of EICO)

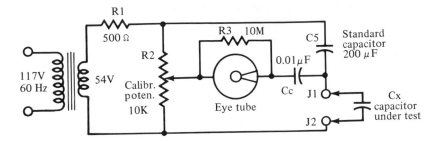

Fig. 5–4 Capacitance-bridge configuration

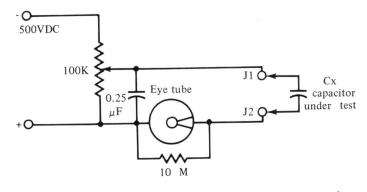

Fig. 5–5 Leakage-indication circuit for paper, mica, and
ceramic capacitors

binding posts on the front panel of the instrument. That is, test leads
must not be employed when measuring small capacitance values because
the stray capacitance of the leads will introduce appreciable error and
cause the indicated capacitance value to be greater than the true value.
Note that 50 volts will appear across J1 and J2 on open circuit. However,
if C_x has a value of 5000 pF, the test voltage is reduced to 2 volts by
capacitive-divider action.

The configuration shown in Fig. 5–4 is employed to measure capacitance values up to 0.5 μF, using a 0.02-μF standard capacitor. A leakage-
indication function is provided for paper, mica, and ceramic capacitors,
utilizing the test circuit shown in Fig. 5–5. A calibrated 100K potentiometer is utilized to apply a potential up to 500 volts across the capacitor
under test. In case the capacitor is leaky, a voltage drop is applied across
the eye tube and the eye sector closes. A capacitor should not be tested
for leakage with a potential greater than its rated working voltage, or
the capacitor is likely to be damaged. Note that the test circuit in Fig.
5–6 is employed to measure capacitance values from 50 μF to 5000 μF.

Any open capacitor will measure zero or near-zero capacitance on the bridge test of Fig. 5–4.

In most procedures, a capacitor is tested for leakage before its capacitance value is measured. If a capacitor has significant leakage, a shallow null is obtained on a capacitance-measurement test. That is, the eye sector opens partially. In case a capacitor has serious leakage, the eye sector remains closed. A capacitor occasionally becomes intermittent. Hence, although a capacitor may appear to be normal otherwise, it is good practice to tap the capacitor during the test. Then, if a mechanical intermittent is present, the opening of the eye sector will fluctuate. If a capacitor is suspected of being thermally intermittent, one of its leads can be heated with a soldering gun during the test.

Electrolytic capacitors are checked with a somewhat elaborated bridge configuration. Thereby, the power factor of the capacitor can be measured, as well as its capacitance value. The configuration shown in Fig. 5–6 is employed to measure capacitance values from 50 μF to 5000 μF, and power-factor values from zero to 80%. This function is utilized chiefly to test electrolytic capacitors. A bridge-driving voltage of 54 volts AC at 60 Hz is used, with no DC polarizing voltage for the electrolytic capacitor under test. Note in passing that some capacitance bridges provide DC polarizing voltage for electrolytic capacitors; this facility permits more precise measurements under simulated load conditions. The AC test voltage applied to C_x in Fig. 5–6 ranges from 50 volts on open circuit to a fraction of a volt when a high value of capacitance is being checked.

To obtain a null in the configuration of Fig. 5–6, both the calibrated potentiometer and the power-factor control must be correctly adjusted. That is, unless the power-factor control is set properly, only a shallow null can be obtained by adjustment of the calibrated potentiometer.

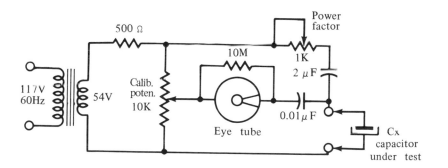

Fig. 5–6   Bridge configuration for checking electrolytic capacitors

Hence, the operator adjusts both controls as required to obtain maximum opening of the eye sector. Note that if an electrolytic capacitor is significantly leaky, an apparently normal null may be obtainable, although the capacitance and power-factor readings will be in error. However, the fact that the capacitor is defective becomes apparent on the leakage test. This function is basically the same as for leakage tests of paper, mica, and ceramic capacitors. Somewhat less sensitivity is provided on the electrolytic-capacitor leakage test because a slight amount of leakage is regarded as normal.

5.3 WHEATSTONE BRIDGE FUNCTION

Many capacitor checkers provide a Wheatstone bridge function for measurement of resistance values. Although not utilized in capacitor checking, a resistance bridge has useful applications in a shop. One of the chief advantages of a resistance bridge is its comparative accuracy and reliability, since its operation does not depend upon batteries, nor is its indication accuracy affected by line-voltage fluctuation. Figure 5–7 shows the configuration for a Wheatstone bridge function of a capacitor checker. We observe that the bridge is energized by AC voltage. The 20-ohm standard resistor provides resistance measurements from 0.5 ohm to 500 ohms. Ranges up to 500 megohms are provided by switching higher values of standard resistors into the circuit.

Note that the bridge shown in Fig. 5–7 is intended only for checking resistors. That is, it is unsuitable for measuring the resistance of an output-transformer primary winding, for example. Since the bridge is energized by AC voltage, a null indication can be obtained only when the component under test has negligible reactance. Since an output transformer has considerable inductive reactance, it draws a lagging current, which upsets the bridge action. Similarly, this type of resistance bridge is unsuitable for measuring the leakage resistance of a capacitor, unless it has a very small capacitance value.

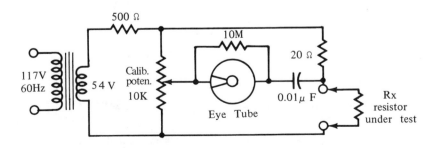

Fig. 5–7 Wheatstone bridge configuration

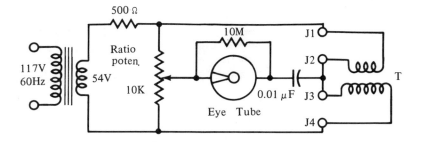

Fig. 5–8   Comparator arrangement for measuring transformer winding ratios

5.4   COMPARATOR FUNCTION

A comparator is a skeleton bridge, as shown in Fig. 5–8. It has two pair of terminals; in this example, the terminals are connected to the primary and secondary windings of a transformer T. A comparator is used chiefly for measuring turns ratios of transformers. We observe that if a transformer has a 1-to-1 turns ratio, that its primary and secondary voltages must be equal. In such case, the bridge balances when the ratio potentiometer is set to its midpoint. The ratio scale is calibrated "1" at its midpoint. Next, if the transformer under test has a 2-to-1 turns ratio, its secondary voltage will be twice its primary voltage. In turn, the bridge balances when the ratio potentiometer is set to a point that provides twice as much voltage to one arm as to the other arm. This point is calibrated "2" on the ratio scale.

A ratio bridge of this type is useful for measuring transformer winding ratios up to 20-to-1 (or 1-to-20). Thus, the ratio scale has a range from 0.05 to 20. The bridge can also be used to determine whether a pair of inductors are identical. That is, if one inductor is connected between J1 and J2, and the other inductor is connected between J3 and J4, the ratio bridge will balance at "1" if the two inductors are identical. However, a simple ratio bridge is inadequate to measure the inductance of an unknown coil with respect to a standard inductor. This inadequacy results from the fact that the resistance values of the two inductors will be unequal, in general, which prevents obtaining a true or even approximate balance.

5.5   COMPLETE CAPACITOR CHECKER

The various test functions that have been described are included in the complete capacitor-checker configuration shown in Fig. 5–9. P1 is the bridge ratio-arm control used on capacitance, resistance, and comparator functions. P2 is the power-factor control, and P3 is the leakage-test voltage

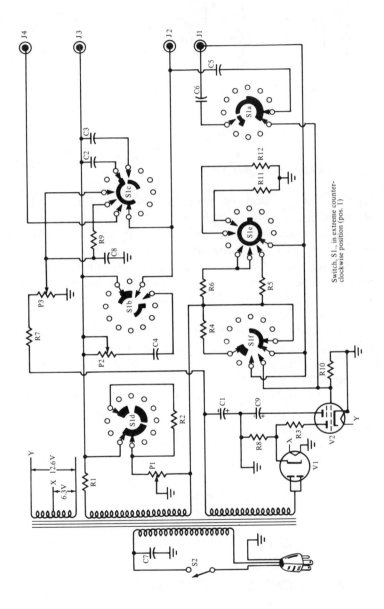

Fig. 5–9 Configuration of a service-type capacitor checker (Courtesy of EICO)

Switch, S1, in extreme counterclockwise position (pos. 1)

122

control. Maintenance of a capacitor checker is usually limited to eventual replacement of the eye tube and rectifier tube. To verify scale calibrations, an assortment of resistors and capacitors with 1% tolerance may be checked on the various ranges. After long service, controls and switches can become worn and require replacement. Electrolytic filter capacitors eventually deteriorate, and fixed paper capacitors may become defective. Fixed resistors are usually extremely long-lived and seldom require replacement.

When making leakage tests of electrolytic capacitors it is essential to observe the polarity markings of the test terminals. In other words, if an attempt is made to check the leakage of an electrolytic capacitor with the test voltage applied in reverse polarity, a good capacitor will appear to be defective. Note also that a good new electrolytic capacitor that has been in storage for a long time may appear to be defective on a leakage test, due to the fact that it requires "forming." This is easily accomplished by connecting the capacitor to the checker, and leaving it on leakage test for an arbitrary period. If the leakage gradually decreases, the capacitor is "forming" and will probably pass a leakage test after a sufficient length of time. "Forming" should be accomplished at rated working voltage of the capacitor.

Note that comparatively few fixed capacitors in radio, TV, and hi-fi equipment have critical capacitance values. That is, the great majority of fixed capacitors have a tolerance of ±20%. Some capacitors will have a tolerance of ±10%. We will find an occasional capacitor with a tolerance rating of ±1%. A close tolerance such as ±1% exceeds the accuracy of a service-type capacitor checker, unless it has been carefully calibrated. Therefore, if a close-tolerance capacitor is suspected of being defective, it should be replaced without attempting to make a definitive test. The same advice applies to capacitors with special temperature coefficients, such as are employed in local-oscillator circuitry. In case of a frequency-drift trouble symptom, temperature-compensating capacitors should be replaced without testing.

5.6   IN-CIRCUIT CAPACITOR CHECKERS

Various types of in-circuit capacitor checkers have been devised, with the objective of facilitating tests of suspected capacitors while shunted by circuit resistance. Although these instruments provide some test information, their limitations tend to outweigh their advantages. In particular, no in-circuit capacitor checker is able to determine whether a capacitor in a solid-state circuit is leaky. Therefore, most technicians rely upon other approaches to the in-circuit capacitor-checking problem. As noted previously, DC voltage measurements provide the best clue to capacitor leakage. Signal-tracing or "bridging" tests provide the best check for suspected open capacitors. In case of remaining doubt, a substitution test is completely conclusive, although it is also the most time-consuming approach.

QUESTIONS AND PROBLEMS

True-False

1. Capacitors give more trouble than other components in radio and television circuitry.
2. A leaky capacitor will show a high resistance reading.
3. An open capacitor will show a high resistance reading.
4. A good 100-pF capacitor will show a high resistance reading.
5. Voltage measurements are a good indicator of an open capacitor.
6. Voltage measurements are a good indicator of a leaky capacitor.
7. A high power factor indicates excessive series resistance.
8. When an eye tube is used to indicate leakage of a capacitor, an open eye sector indicates a shorted capacitor.
9. A slight amount of leakage is considered normal in an electrolytic capacitor.
10. A ratio bridge is useful in measuring leakage of capacitors with capacitance values as low as 5 pF.
11. Capacitor checkers require calibration and maintenance often.
12. An electrolytic capacitor that has been stored for a long period will often show excessive leakage.
13. Capacitors with close tolerance should be replaced if they are suspected of being the cause of a problem.
14. DC voltage measurements give the best clues for in-circuit capacitor testing.

Multiple-Choice

1. The simplest way to test for an open capacitor is
   (a) with a capacitor tester.
   (b) with an ohmmeter.
   (c) by bridging a good capacitor across the doubtful unit.
2. A leaky capacitor can usually be found by
   (a) the replacement method.
   (b) bridging the capacitor with a good unit.
   (c) a resistance test with an ohmmeter.

3. The resistance reading of a 100-pF capacitor will be
   (a) low.
   (b) high.
   (c) low at first and slowly rise.
4. Voltage measurements cannot be used to locate a
   (a) shorted capacitor.
   (b) leaky capacitor.
   (c) open capacitor.
5. The impedance of a capacitor can be checked with a(n)
   (a) ohmmeter reading.
   (b) power factor indicator.
   (c) voltmeter.
6. The power factor of an ideal capacitor is
   (a) high.
   (b) about 10.
   (c) zero.
7. A comparator is used chiefly to measure
   (a) capacitance of a transformer.
   (b) leakage of a capacitor.
   (c) the turns ratio of a transformer.
8. An electrolytic capacitor that has been stored for a long period often shows excessive leakage. To correct the problem you should
   (a) heat the leads.
   (b) form the capacitor with a voltage.
   (c) throw out the capacitor.
9. A capacitor tester will not check
   (a) the value of the capacitor.
   (b) the temperature coefficient of a capacitor.
   (c) leakage.

General

1. What is the easiest method of testing a capacitor that is suspected of being open?
2. What is the most common capacitor problem?
3. What is the purpose of the power factor function on a capacitor tester?
4. What functions do most service-type capacitor checkers provide?
5. How is a thermal intermittent located in a capacitor?
6. How are leaky capacitors checked in semiconductor circuits?

6

Signal Tracers
and
Analyzers

6.1 SIGNAL-TRACER AND ANALYZER
REQUIREMENTS

Radio and television receivers are signal processors, directing an incoming signal voltage along a channel or channels from the antenna-input terminals to the speaker and/or the picture tube. Figure 6–1 shows the signal channel and processing sequence for a conventional AM broadcast radio receiver. We observe that there are six basic sections in which the signal could become distorted, attenuated, or completely stopped. Preliminary trouble-shooting procedures are concerned with sectionalization of the receiver defect. We observe that if a suitable signal indicator is applied step-by-step through the signal channel, the presence or absence of signal voltage can be determined at each section. If the indicator includes a meter or equivalent device, the amplitude of the signal voltage can be measured in each section. Again, the indicator can be designed to show whether the signal is distorted or normal.

Figure 6–2 shows the signal channels and processing sequence for a black-and-white television receiver. Note that the basic arrangement is much the same as in a radio receiver, although the incoming signal is more complex and more sections are utilized. It is evident that the sweep waveforms are not signal voltages, in that these waveforms are normally present whether an antenna-input signal is applied or not. On the other hand, the sync waveform is a signal voltage, inasmuch as the sync waveform is not present unless an antenna-input signal is applied. Both radio and television receivers employ power supplies, which are not a part of the signal channels. Again, radio receivers have an AVC (automatic volume control) section, and TV receivers have an automatic gain control (AGC) section, which are not a part of the signal channels.

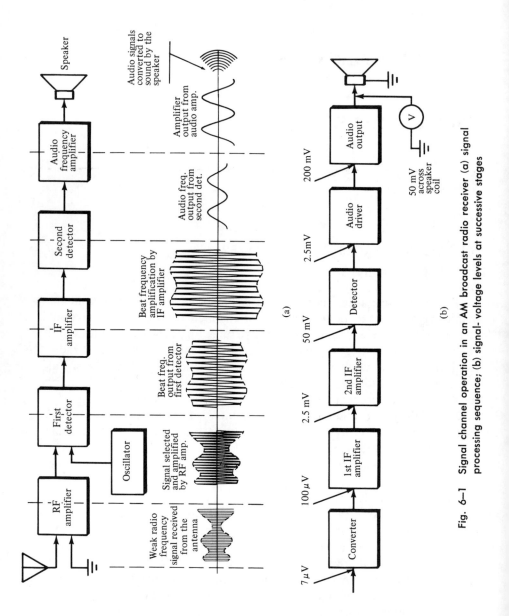

Fig. 6–1 Signal channel operation in an AM broadcast radio receiver (a) signal processing sequence; (b) signal-voltage levels at successive stages

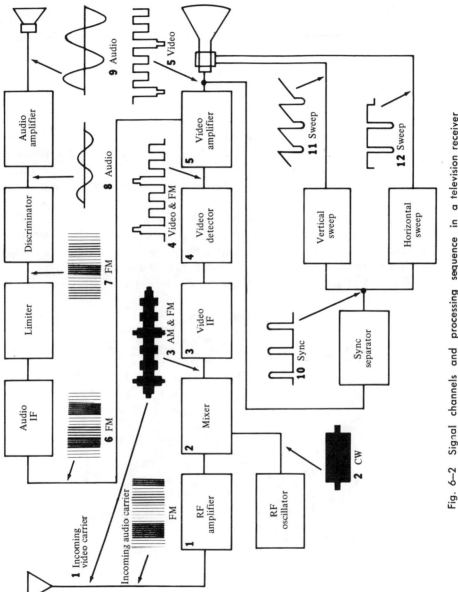

Fig. 6–2 Signal channels and processing sequence in a television receiver

6.2   SIGNAL TRACING AND SIGNAL SUBSTITUTION

Signal-tracing procedures start at the antenna-input terminals of a receiver, and progress section by section to the speaker, picture tube, or sweep sections. Either an incoming broadcast-station signal may be utilized or a generator signal may be applied to the antenna-input terminals. If a generator signal is utilized, an amplitude-modulated RF waveform is applied to the input of an AM radio receiver. A sweep signal (FM waveform) is preferred when an FM radio receiver is under test. In the case of a black-and-white TV receiver, a test-pattern generator signal provides the most useful signal information. Or, when a color-TV receiver is under test, a color-bar signal is required to energize the chroma section of the receiver.

Indicator units or devices used in signal-tracing procedures should be appropriate for the circuits under test. As noted previously, a sensitive oscilloscope is the most generally applicable indicating instrument, when supplemented by low-capacitance and demodulator probes. Similarly, a sensitive TVM is very useful, if supplemented by an RF probe. Radio-

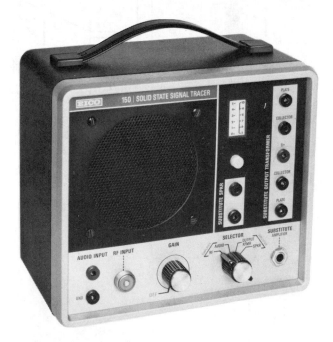

Fig. 6–3   Specialized radio-receiver signal tracer (Courtesy of EICO)

Fig. 6–4 Appearance of television analyst, a specialized type of television signal generator (Courtesy of B & K Mfg. Co.)

receiver signal tracing is often accomplished by specialized signal-tracing instruments, such as illustrated in Fig. 6–3. This instrument provides both audible indication and a meter measurement of signal strength. As explained in greater detail subsequently, this type of signal tracer checks RF, IF, and AF receiver sections when used with supplementary probes. Although a radio signal tracer has some degree of usefulness in TV servicing procedures, we will find that the oscilloscope provides the most useful signal information.

Next, it is instructive to note the general principles of signal-substitution test procedures. This method of sectionalization employs the speaker and/or picture tube as an indicator. Generator signal voltages are applied to each stage of the receiver to determine whether the stage will pass the signal, or if the signal is stopped or weakened. If a radio receiver is under test, an AF signal is applied at the speaker terminals, and if an audio tone is audible, the AF signal is applied next at the input of the audio-output stage. The IF stages are tested with a modulated IF signal; an amplitude-modulated signal is employed in the case of an AM receiver, but a frequency-modulated (sweep) signal is utilized to better advantage in the case of an FM receiver. Similarly, an AM or FM RF signal is used to test the front end of the receiver. In the case of a TV receiver, it is advisable to apply appropriate signals from an analyst type of generator, as illustrated in Fig. 6–4. TV analyzers are explained in greater detail later.

6.3   OPERATION OF A RADIO
SIGNAL TRACER

Specialized radio signal tracers always provide speaker output. Many also provide meter indication (Fig. 6–5). Two probes are employed, so that either high-frequency or low-frequency signals can be checked. Low-frequency signals are checked with the direct probe which feeds the signal into the audio amplifier. A gain control is provided so that the amplifier will not be overloaded on high-level signals. Both a speaker and a meter are driven by the amplifier. When signal frequencies higher than the audio range are to be checked, the RF probe is utilized. This probe operates as a traveling detector, and demodulates the signal, as shown in Fig. 6–6. Thereby, an audio-frequency output is produced by an RF probe.

Figure 6–7 shows the configuration for a typical radio signal tracer. It comprises a high-gain audio amplifier with speaker and meter output indicators. An RF probe and a direct (audio) probe are provided. A supplementary built-in output transformer is also included. This transformer can be used with the internal speaker for tests of either single-ended or

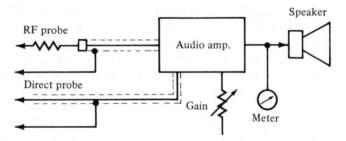

Fig. 6–5   Plan of a radio signal tracer

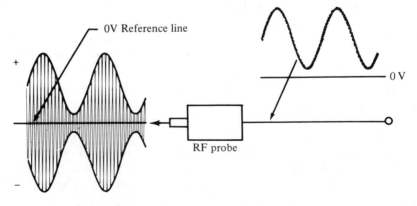

Fig. 6–6   Signal processing action of an RF probe

push-pull audio amplifiers. Both solid-state and tube-type amplifiers are accommodated by the tapped primary winding. The switching arrangement also permits the internal speaker to be used as a substitute speaker for amplifiers or audio equipment under test. A jack is provided, whereby the signal-tracer amplifier can be utilized as a substitute amplifier.

Although it is advantageous to use an AM generator signal in weak-signal areas and in stage-gain estimations, a broadcast-station signal is more useful in checking distortion. When signal-tracing FM receivers or the sound section of a TV receiver, an AM signal generator can be used to check the stages up to the FM detector. Since service-type AM generators often have residual FM in their AM output (particularly when operated at a high percentage of modulation), it is often possible to also check the FM detector and the following audio section. Otherwise, an FM generator such as a sweep-frequency generator should be used to energize FM circuitry. A sweep-frequency generator provides a 60-Hz tone.

Approximate gain or loss measurements can be made by observing the pointer deflection on the meter as the test point is moved from the input to the output stage. In case the pointer is deflected off-scale, the setting of the gain control is reduced; the amount of change required in gain-control setting to bring the pointer back to its original indication then becomes a check on stage gain. These gain measurements are approximate, particularly when using the RF probe, because the probe tends to load and/or detune high-frequency circuitry. In some cases, a stage under test might become sufficiently disturbed that it will oscillate when the RF probe is applied. An oscillating stage appears to be "dead." Note that if a normal signal is found at the following test point, it will be concluded that the circuit disturbance caused the previous "dead" indication.

Audio stages operate at a comparatively high signal level. For this reason, the signal tracer is operated at reduced sensitivity in audio tests. With reference to Fig. 6–7, capacitor C5 is disconnected from ground when the audio probe is switched into use. In turn, degeneration in the Q2 circuitry results in reduction of indication sensitivity. When signal-tracing through an audio amplifier, it will be observed that there is normally a substantial reduction in signal level from the primary to the secondary of the output transformer. This step-down in signal voltage is due to the impedance relations provided by the output transformer. That is, the transformer changes the comparatively high output impedance of the amplifier to a low impedance that matches the speaker input impedance. In turn, the signal voltage is considerably reduced, although the signal current is considerably increased.

We will find that there are certain points in receiver circuitry where the presence of signal indicates trouble. For example, let us consider

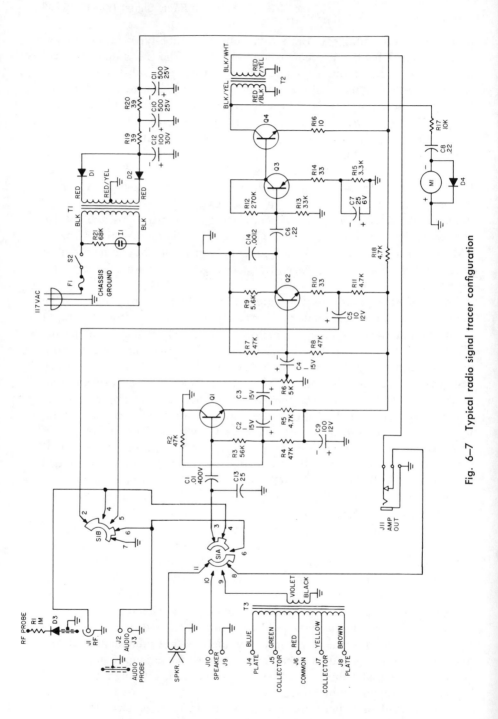

Fig. 6-7 Typical radio signal tracer configuration

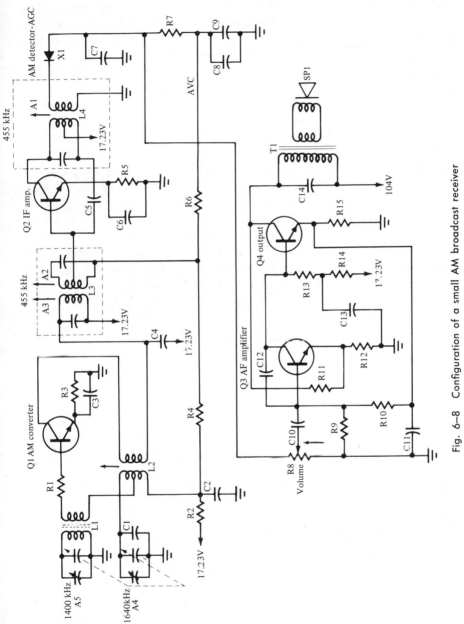

Fig. 6-8 Configuration of a small AM broadcast receiver

135

capacitor C4 in Fig. 6–8. This is an emitter-bypass capacitor for Q4, and a base-decoupling capacitor for Q3. There is normally no measurable signal voltage across C4. On the other hand, if C4 loses a substantial portion of its rated capacitance value, trouble symptoms will develop and we will find a signal voltage present across C4. Similarly, practically no signal voltage is normally present across C2 or C5. C2 is an AVC bypass capacitor, and C5 is a base-decoupling capacitor for Q4. The same observations apply to C14 and C16. These are emitter-bypass capacitors, and an open capacitor will result in considerable reduction of stage gain.

With reference to Fig. 6–7, the multi-tap transformer has its secondary winding connected to the internal speaker in the corresponding switch position. Terminals for the primary winding are available at five pin jacks on the instrument panel. This output transformer is designed for use with single-ended or push-pull amplifiers that employ either transistors or tubes. Thus, substitution tests can be made of suspected defective output transformers. Several primary impedances are provided, so that a practical match can be made to the amplifier under test. The internal speaker can also be switched over to a pair of front-panel terminals, so that it can be used in substitution tests of suspected defective speakers.

6.4   OPERATION OF A TELEVISION ANALYZER

As noted previously, a television analyzer is a specialized type of signal generator, and is particularly adapted for signal-substitution tests in television receivers. Figure 6–9 shows a block diagram for a TV analyzer designed for troubleshooting both black-and-white and color-TV receivers. We observe that 10 oscillators are employed, comprising UHF, VHF, IF, vertical, horizontal, sound (1 kHz), intercarrier (4.5 MHz), color (3.563 MHz), bar (189 kHz), and shorted-turns oscillators. In addition a test-pattern generator is provided. This unit consists of a flying-spot scanner, which forms a video signal from a slide transparency. The essential components in the flying-spot arrangement are a cathode-ray tube and a photomultiplier tube.

A television analyzer is utilized to inject appropriate signal voltages at various points in the signal channels. The picture tube and/or the speaker of the receiver under test are employed as indicators. When the UHF, VHF, or IF oscillators are used, a test-pattern video signal is amplitude-modulated on the CW output from the oscillator. In turn, a test pattern such as illustrated in Fig. 6–10, will be displayed on the picture-tube screen if the receiver is operating normally. Considerable information

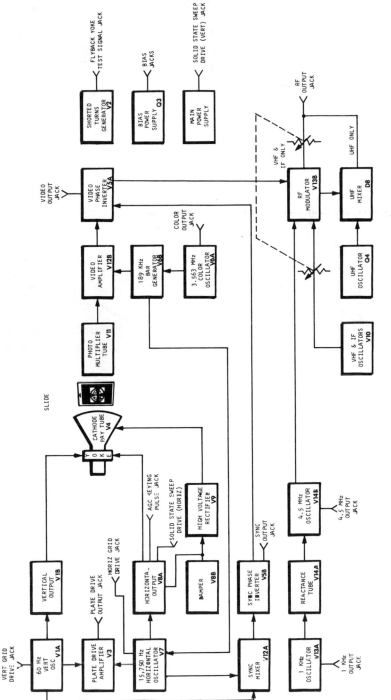

Fig. 6–9 Block diagram of a television analyzer (Courtesy of B & K Mfg. Co.)

137

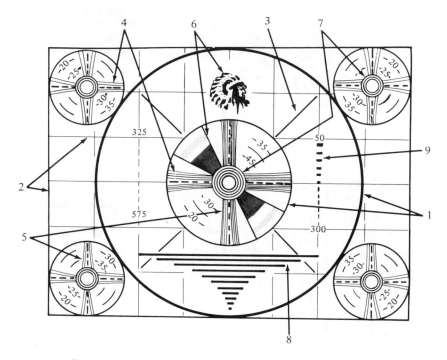

Fig. 6–10 Standard test pattern. (See text for explanation of numbered features.)

concerning receiver circuit action is provided by a test pattern. Various distortions, numbered in Fig. 6–10 as here, are evaluated as follows:

1. Height, width, linearity, and centering adjustments are indicated by the shapes of the circles. All circles are normally round, within practical tolerances, and not more than ¾ inch is cut off from the circles by the edges of the screen.

2. Aspect ratio, pincushion distortion, or barrel distortion are shown by the vertical and horizontal lines, which normally form squares. True squares correspond to the standard 4:3 aspect ratio. These lines are normally straight; if curved, the presence of pincushion or barrel distortion is indicated, as exemplified in Fig. 6–11. Again, if the lines are not parallel, keystoning is present, as depicted in Fig. 6–12. Or, if the lines are parallel but unevenly spaced, nonlinear deflection is occurring, as shown in Fig. 6–13.

3. Poor interlacing is indicated by a jagged or sawtooth display of the normally straight diagonal lines. All four lines are affected by interlacing faults.

4. Vertical resolution is indicated by the horizontal wedges in the test

pattern. The wedges are normally clear and sharply defined, but become indistinct and "washed out" when the vertical resolution is subnormal. Note that the horizontal wedges will also be affected by poor interlacing.

5. Signal-channel bandwidth is indicated by the vertical wedges. We observe the number along the wedge at which the vertical wedge becomes indistinct, and add a zero. This gives the horizontal resolution in number of vertical lines; this number divided by 80 gives the signal channel bandwidth in MHz.

6. Picture-signal and picture-tube linearity (gamma) is indicated by the

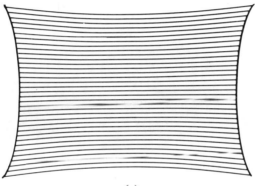

(a)

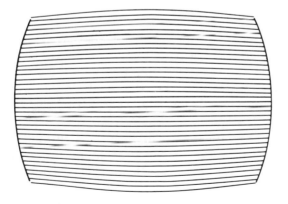

(b)

Fig. 6–11   Curvature of normally straight lines in test pattern:
(a) pincushion distortion; (b) barrel distortion

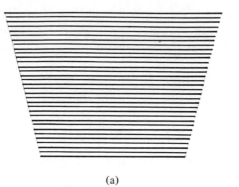

(a)

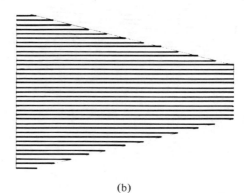

(b)

Fig. 6–12   Examples of keystoned rasters: (a) horizontal keystoning; (b) vertical keystoning

diagonal wedges. Nonuniform contrast gradation in wedge display indicates poor gamma. These wedges are also useful in adjusting the contrast, brightness, and AGC controls. Four shading tones are provided in the wedges, in steps from black through gray to white.

7. Focus of the picture-tube beam is indicated by the innermost circles or "bulls-eye."

8. Frequency response of the signal channel is also indicated by the 11 horizontal bars of various lengths. From top to bottom, these bars correspond to square-wave frequencies of 19 kHz, 28 kHz, 38 kHz, 56 kHz, 75 kHz, 113 kHz, 150 kHz, 225 kHz, 300 kHz, 450 kHz, and 600 kHz. Poor low-frequency response shows up as disappearance or blurry reproduction of various bars.

9. Horizontal resolution is indicated by the single resolution lines. From top to bottom, these lines progress in steps of 25, ranging from 50 to

575 lines of horizontal resolution. This corresponds to a frequency range from 600 kHz to 7 MHz. Disappearance or blurry reproduction of various lines indicates faulty horizontal resolution. To obtain the frequency corresponding to a line, divide the line number by 80.

When it appears that the receiver is not operating normally, or if no picture is displayed, test signals are then injected at various points in the signal channel to localize the defective section. For example, we may proceed to check at the input of the IF amplifier by injecting an IF test-pattern signal at the base of the third IF transistor (Fig. 6–14). In case a

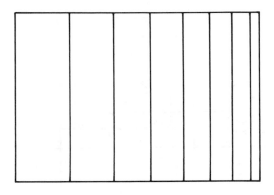

(a)

(b)

Fig. 6–13 Two types of scanning nonlinearity: (a) horizontal nonlinearity, right-side cramping; (b) vertical nonlinearity, bottom cramping

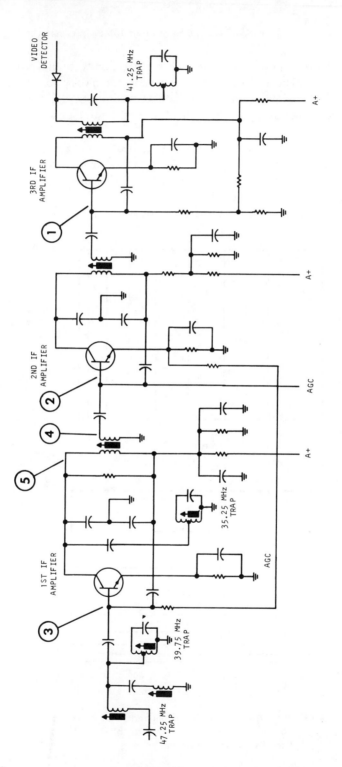

Fig. 6–14   Progressive signal-injection points in an IF-amplifier strip (Courtesy of B & K Mfg. Co.)

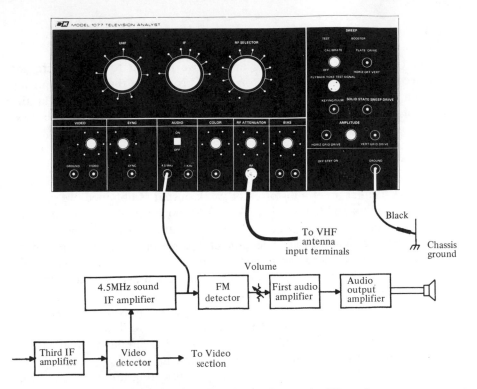

Fig. 6–15   Signal substitution in the intercarrier-FM section
(Courtesy of B & K Mfg. Co.)

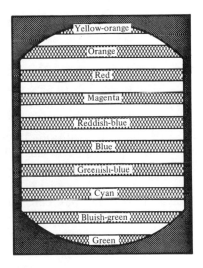

Fig. 6–16   Display of keyed-rainbow color test signal on
screen of color picture tube (Courtesy of B & K
Mfg. Co.)

normal pattern display does not appear, we proceed to the collector of the second IF transistor, and so on. If we proceed to the video-frequency section of the receiver, we may inject a VF test-pattern signal at the base of the first video-amplifier transistor, and then proceed, if necessary, to the grid of the picture tube.

In case the intercarrier FM-IF section falls under suspicion, a 4.5-MHz test signal modulated by a 1-kHz audio signal can be injected stage by stage, as depicted in Fig. 6–15. In case no sound output is obtained from the speaker, or if a distorted or weak output is obtained, we proceed into the audio-amplifier section. A 1-kHz audio signal is used to check the audio stages and the speaker. Attenuators are provided in the analyzer for control of signal levels. The gain of a stage can be approximated by noting how far an attenuator must be turned back to obtain the same signal output when the signal-injection point is moved ahead from the base to the collector of a transistor.

Sync-section troubles are analyzed by injecting a sync signal stage-by stage to determine the point where the test pattern will be locked in sync instead of tearing horizontally or rolling vertically. Similarly, AGC-section defects are analyzed by injecting an AGC keying pulse at suitable points. Again, sweep-section faults are analyzed by injecting horizontal-sweep or vertical-sweep waveform voltages into the corresponding circuits. A high-voltage test facility is provided, in addition to an inductor test arrangement, to determine whether shorted turns are present in a deflection yoke or flyback transformer.

Chroma circuitry in a color-TV receiver is checked by injecting a keyed-rainbow test signal at successive stages in the chroma section. A color test pattern, such as shown in Fig. 6–16, is normally displayed on the screen of the color picture tube. Signal injection may start at either the input or the output of the chroma section. In case of a no-color symptom, with black-and-white reception normal, we may start the signal-injection procedure at the input of the chroma demodulators, as outlined in Fig. 6–17. Or, if there is a loss of one color, with others normally reproduced, we may start the signal-injection procedure at the color picture tube, and work back step by step to the chroma demodulators, as outlined in Fig. 6–18.

In conclusion, a television analyzer has considerable versatility and can be operated advantageously by personnel with a minimum knowledge of electronic instruments. Since the picture tube and the speaker are utilized as indicators, no interpretation of scale readings or waveforms is required. The operator needs to know only the pertinent test points in the receiver chassis.

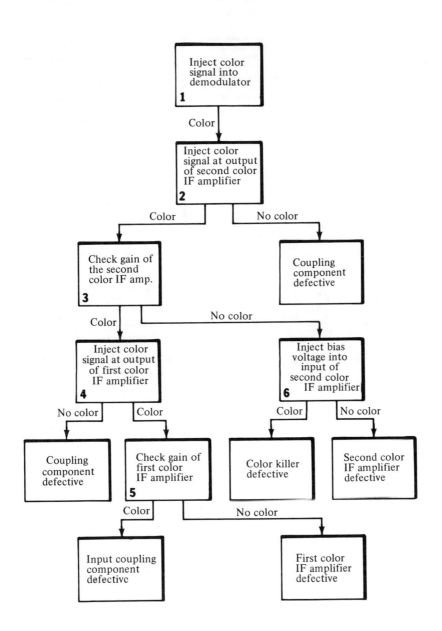

Fig. 6–17   Signal-injection procedure for a no-color trouble
symptom (Courtesy of B & K Mfg. Co.)

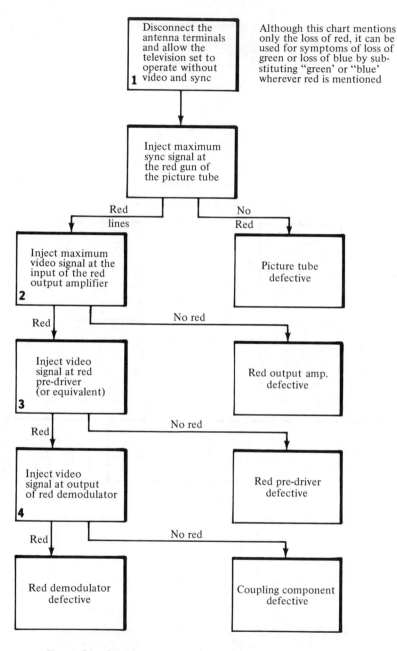

Fig. 6–18 Signal-injection procedure for a one-color-absent symptom (Courtesy of B & K Mfg. Co.)

QUESTIONS AND PROBLEMS

True-False

1. In order to troubleshoot a radio or television, we should know the signal processing path.
2. Power supplies are part of the signal channels.
3. The oscilloscope provides the most useful signal information when signal tracing a TV receiver.
4. The speaker and picture tube can be used in the signal-substitution test procedure.
5. When using a radio signal tracer it is necessary to use an AM signal generator for a signal source.
6. A broadcast station is more useful for checking receiver distortion than an AM generator.
7. It is often possible to check the FM detector of a TV receiver with an AM generator.
8. An RF probe can cause a stage to oscillate and appear dead.
9. Audio stages in a radio operate at low signal levels.
10. A signal generator is necessary when making gain measurements of each stage in a radio.
11. The RF probe tends to load or detune high-frequency circuitry.
12. There is normally an increase in voltage level from the primary to the secondary of the output transformer.
13. An open emitter bypass capacitor will result in decreased stage gain.
14. A television analyzer is a special type of signal generator.
15. A television analyzer is a special type of signal generator that uses two oscillators.
16. A television analyzer uses the picture tube or speaker as indicators.
17. In reference to Fig. 6–13, the pattern shows the results of the keystone effect.
18. Poor interlacing is indicated by jagged or sawtooth display of the normally straight diagonal lines.
19. The television analyzer can be used as a signal tracer up to and including the input to the picture tube control grid.
20. In case of a no-color symptom, with black-and-white reception normal, we may start the injection procedure at the input of the chroma demodulators.

Multiple-Choice

1. The _____ is part of the signal processing channel of a TV receiver.
   (a) RF amplifier
   (b) power supply
   (c) horizontal oscillator

2. When an FM receiver is being tested by signal substitution, it is preferred to use a(n) _____ generator.
   (a) AM
   (b) FM
   (c) sweep-frequency

3. A radio signal tracer is usually not supplied with
   (a) an internal signal source.
   (b) a speaker.
   (c) an RF probe.

4. The purpose of using an RF probe with a signal tracer is to
   (a) prevent oscillation of the stage.
   (b) convert the RF signal to an audio signal.
   (c) prevent loading of an RF amplifier.

5. The voltage at the secondary of the output transformer is less than the voltage at the input because the
   (a) transformer has a step-up turns ratio.
   (b) transformer has a step-down turns ratio.
   (c) speaker loads the circuit.

6. An open emitter bypass capacitor will result in
   (a) reduction in gain of the stage.
   (b) a change in the bias of the stage.
   (c) distortion.

7. An instrument that employs several oscillators and is used in trouble-shooting both black-and-white and color TV is called
   (a) an oscilloscope.
   (b) a sweep generator.
   (c) a TV analyzer.

8. When a TV analyzer is used the indicator is
   (a) a TVM.
   (b) an oscilloscope.
   (c) the picture tube or speaker.

9. Disappearance or blurry reproduction of various lines, indicated in Fig. 6–10, is the result of
   (a) faulty horizontal resolution.
   (b) poor focus.
   (c) vertical nonlinearity.

10. The television analyzer is basically a
    (a) simple signal generator.
    (b) television receiver substitution generator.
    (c) video signal injector.

General

1. Basically, what are radio and television receivers?
2. Where should you start signal tracing a receiver?
3. What is the most useful indicating instrument?
4. What is a signal substitutor?

5. How are both high-frequency and low-frequency signals checked with a signal tracer?

6. What are the basic blocks of a typical signal tracer?

7. Why is it that AM signal generators can be used to test FM receivers?

8. What is a problem in using an RF probe?

9. Why is there normally some decrease in signal level between the input and the output of an output transformer?

10. What is the purpose of a bypass capacitor?

11. How is the speaker of a radio receiver tested with a signal tracer?

12. What is the function of the flying-spot scanner?

13. What would you use as an indicator if you were testing a television receiver with a flying-spot scanner?

14. What are some of the distortions that might show on the picture tube?

15. How is a defective section of a television located?

16. How are signal levels controlled in an analyzer?

17. How is the chroma section of a color-TV receiver tested with a TV analyzer?

7

Transistor
Testers

7.1 TRANSISTOR TESTER REQUIREMENTS

Although there are many transistor parameters that are measured by engineers in laboratories, service-type transistor testers usually measure the beta value and the collector leakage current of the transistor under test. Some instruments are designed to check transistors out of circuit only, while others also provide in-circuit tests. In-circuit tests are necessarily limited, inasmuch as the transistor must be checked while connected to arbitrary values of resistance, capacitance, and inductance; or the transistor under test might be connected to another transistor, or to one or more diodes. Most service-type instruments provide measurements of DC beta. That is, the transistor is checked in the common-emitter configuration; a change in DC base current is applied, and the corresponding change in DC collector current is measured.

Leakage is generally measured from collector to base, with the emitter terminal open (I_{CBO}), or from collector to emitter, with the base terminal open (I_{CEO}). Diodes are checked for front-to-back ratio. Note that transistors and diodes employed in integrated circuits cannot be checked with service-type testers. It is not possible to check a transistor for I_{CBO} or I_{CEO} while it is connected into a circuit. However, it is usually possible to determine whether the transistor has an effective DC beta. The chief exception occurs when a transistor is connected across a low-resistance component, such as a coil. In-circuit tests are not quantitative, but are often useful as a go/no-go form of quick check.

7.2   CONSTRUCTION AND OPERATING
FEATURES OF AN OUT-OF-CIRCUIT
TRANSISTOR TESTER

A typical service-type transistor tester for out-of-circuit applications is illustrated in Fig. 7–1. It utilizes self-contained batteries and is fully

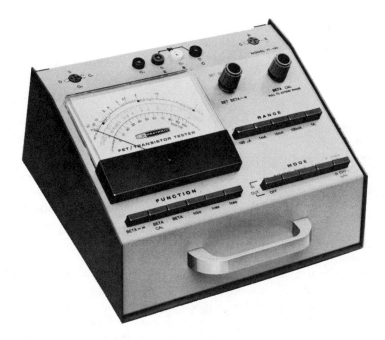

Fig. 7–1   Appearance of an out-of-circuit transistor tester
(Courtesy of Heath Co.)

portable. Beta values up to 400 are indicated directly, with adjustable test voltages so that measurements are made under the rated operating conditions for a transistor. It is instructive to observe each test function that is provided in this example. Although we will consider tests of a PNT-type transistor, tests for an NPN transistor are exactly the same except for reversed polarities of the electrode potentials. Figure 7–2 depicts the base-current test (I_B). Note that the indicating meter is connected in series with the base circuit. In turn, the base current is measured. The value of base current is adjusted by means of the bias control R6. Collector voltage is adjustable with a switch (not shown). R_s is a meter shunt resistor. Control R6 is adjusted for a specified value of base current. The resulting collector current is then measured, as explained next.

Let us consider the transistor gain test shown in Fig. 7–3. This is a

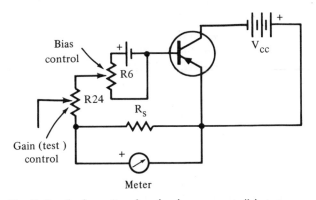

Fig. 7–2 Configuration for the base-current (I_B) test

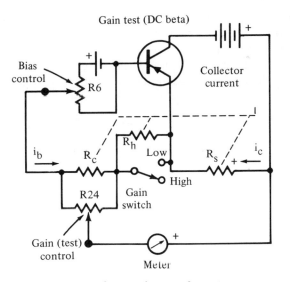

Fig. 7–3 Gain test (beta or h_{FE}) configuration

DC beta or h_{FE} measurement. The ratio between collector current and base current is indicated by the front-panel gain control on a scale calibrated in DC beta values. We observe in Fig. 7–3 that the base current produces a voltage drop across the gain test control R24. In turn, the collector current produces a voltage drop across R_s. Note that the meter is connected from the collector end of R_s to the arms of R24. With the gain control R24 initially set to its right-hand end, the meter indicates the collector current value. Then, as the arm of the gain control is moved toward its left-hand end, an equipotential point is found at which the meter indicates zero current. At this point, the ratio of collector current to base current is indicated by the pointer on the gain-control scale. That is, we null the meter and then read the DC beta value directly.

This gain-control scale is calibrated from 0 to 200, and from 200 to 400. If a transistor under test happens to have a beta value greater than 200, the gain switch is turned to its High position. In turn, resistor R_H (Fig. 7–3) is switched in series with R24, thereby extending the measurement range from 200 to 400. Tests of high-current transistors require that R_S be changed to a smaller value. In turn, the values of R_H and R_C must be changed in the same proportion, so that the base and collector current ratio remains constant. These changes in resistance values are made by turning the collector-current switch.

Figure 7–4 shows the configuration that is used in the collector-voltage (E_C) measurement. A common-emitter arrangement is utilized. The meter indicates the collector-emitter voltage. This voltage is adjustable by means of the collector-voltage switch (not shown). Resistor R_V is a meter multiplier. Note that the collector voltage is measured at a specified value of base current which is determined by the setting of bias control R6.

Next, the collector-current (I_C) measurement configuration is depicted in Fig. 7–5. The transistor is connected in the CE mode, and the current meter is placed in the collector-emitter circuit. Note that the collector

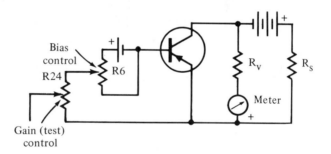

Fig. 7–4   Collector-voltage (E_c) measurement

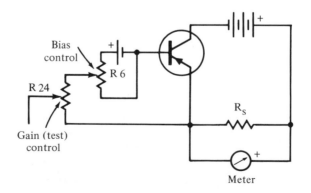

Fig. 7–5   Collector-current (I_c) measurement

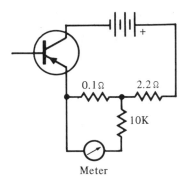

Fig. 7–6   Transistor short-circuit test

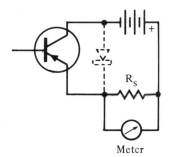

Fig. 7–7   Collector-to-emitter leakage (I_{CEO}) or diode test

current is measured with a specified value of base current which is determined by the setting of bias control R6.

Transistors sometimes become defective because of a short-circuit from collector to base, and a "short" test is provided as shown in Fig. 7–6. We observe that the voltage selected by the collector-voltage switch is applied to the collector and emitter, first through a 0.1-ohm and then through a 2.2-ohm resistor. The meter with its 10K dropping resistor is connected across the 0.1-ohm resistor. If the transistor is not short-circuited, the meter indicates zero. On the other hand, a "dead" short in the transistor will produce full-scale deflection on the meter.

Next, let us consider the collector-to-emitter leakage (I_{CEO}) or diode test depicted in Fig. 7–7. A specified test voltage is applied between collector and emitter; in turn, the leakage current with the base open is indicated by the meter. Note that if a diode (dashed lines) is being checked for reverse current flow, it is connected in place of the transistor with the polarity indicated. To measure the forward current flow of the diode, the polarity of the test voltage is reversed.

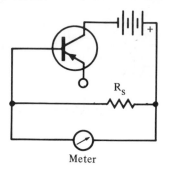

Fig. 7–8 Collector-to-base leakage (I_{CBO}) test

Figure 7–8 shows the configuration that is used in the collector-to-base leakage (I_{CBO}) test. The meter indicates the leakage current between collector and base, with the emitter open. If a transistor is in normal operating condition, the leakage current in an I_{CEO} test will be approximately beta times the leakage current in an I_{CBO} test. This relationship exists because the base is not connected to the emitter in the I_{CEO} test. With the base electrode open, or "floating," the voltage drop produced by the leakage current across the base-emitter junction resistance serves as a small bias which increases the collector current flow accordingly.

7.3 CONSTRUCTION AND OPERATION OF AN IN-CIRCUIT TRANSISTOR TESTER

A typical in-circuit/out-of-circuit transistor tester is illustrated in Fig. 7–9. The corresponding instrument configuration is shown in Fig. 7–10. Although a simplified design is employed, the operation of the tester is essentially the same as explained for the preceding out-of-circuit instrument. All tests are made with a 1.5-volt DC source. First, the beta-calibration control, R3, is adjusted to provide full-scale indication with the beta test switch in its Calibrate position. This adjustment provides a reference value of forward-current flow through the base-emitter junction of the transistor under test. Next, the beta test switch is thrown to its Test position. This places the meter in the collector circuit of the transistor, and the DC beta value is read directly on the meter scale.

Out-of-circuit beta measurements are quite accurate in the foregoing test. However, the accuracy of an in-circuit test depends on the shunting resistances of the circuit into which the transistor is connected. Therefore, if any gain is indicated by the meter, the transistor is considered to be good on an in-circuit test. Note that if the transistor is an NPN type and a test is attempted for a PNP type, the meter will not deflect to the Calibrate point. Similarly, a PNP transistor will not produce a pointer deflection to the Calibrate point, if the instrument is set for an NPN test. However, if the meter will not calibrate normally in either polarity, the transistor may be open-circuited or short-circuited. Note that if the collector and emitter test leads are accidentally interchanged, the apparent gain of the transistor will be subnormal.

We will observe that the ability of the in-circuit test to check the gain of a transistor in-circuit is based on the low resistance of the instrument circuitry. With reference to Fig. 7–10, we observe that the resistance between the green and white leads may be as low as 25 ohms, and the resistance between the red and green leads is approximately 12 ohms. In turn, the shunting resistances of the circuit into which the transistor is connected are often very much greater than the instrument internal resistances. However, these shunting resistances may be of the same order of magnitude, or even less than the instrument internal resistances in some situations. In turn, an in-circuit gain check becomes impractical.

Leakage tests must always be made with the transistor out of the circuit because even a high value of shunting resistance will give a misleading

Fig. 7–9   Appearance of an in-circuit/out-of-circuit transistor
tester (Courtesy of Heath Co.)

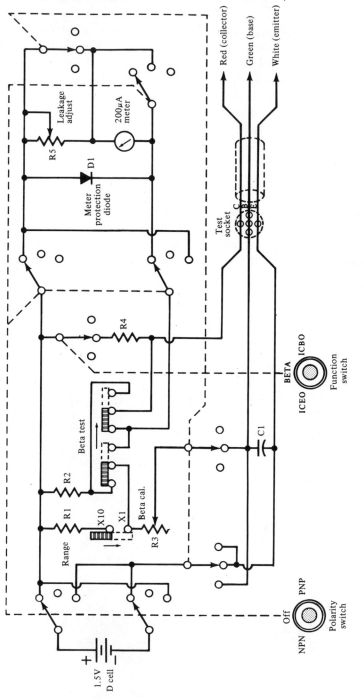

Fig. 7-10 In-circuit/out-of-circuit transistor-tester configuration

indication of leakage current. As a practical operating note, technicians sometimes disconnect a transistor from a printed circuit without unsoldering the leads to the transistor. To do this, a razor blade or other sharp-edged device is used to make a thin cut across the printed conductors to the base and emitter. Only two of the conductors need to be opened for an effective out-of-circuit test. Then, if the test results indicate that the transistor is not defective, the circuit is restored by melting a small drop of solder across the cut in each of the PC conductors.

7.4 TRANSISTOR CURVE TRACER

Another type of transistor tester used in service work is depicted in Fig. 7–11. This arrangement is called a transistor curve tracer, and the test data are displayed as a collector family of characteristic curves on the screen of an oscilloscope. That is, the relation between collector voltage and collector current for several values of base-bias voltage is displayed as in Fig. 7–12. The amplitude of the pattern and the slope of the characteristics are the chief features of the test information. We observe in Fig. 7–11 that the base of the transistor under test is biased by a staircase-voltage waveform. In turn, the forward-current flow into the base is changed by discrete values. From three to six steps are employed by typical service-type testers. In this example, three steps are shown, each with a duration of 1/120 second.

Next, let us consider how the CRT display is developed in Fig. 7–11. The collector source voltage is a 120-Hz half-sine waveform. However, the exact waveform is inconsequential because Lissajous-type traces are utilized. This half-sine voltage varies from zero to the maximum rated collector voltage for the transistor under test. Note that the collector current flows through a 50-ohm resistor. In turn, the voltage drop across the resistor is proportional to collector-current flow. Therefore, the

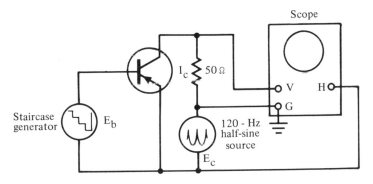

Fig. 7–11 Basic curve-tracer arrangement

vertical deflection represents collector current, and the horizontal deflection essentially represents collector voltage on the CRT screen. Each time that the base current is stepped up, a Lissajous trace is displayed at a higher level on the CRT screen.

It is evident that if the collector junction of the transistor is short-circuited in Fig. 7–11, the display will consist of a nearly vertical trace because the vertical amplifier is operated at high gain. A high-gain setting is employed because the voltage drop across the 50-ohm resistor is quite small. On the other hand, if the collector junction is open-circuited, no current flows through the 50-ohm resistor, and the display will consist of a horizontal trace. Collector leakage is equivalent to connecting a resistor between the collector and emitter terminals of the transistor. The result is to produce excessive collector current flows and to expand the pattern to an abnormal height. Most service-type curve tracers can be calibrated, so that the levels and slopes of the traces can be compared with specified collector families, as shown in Fig. 7–12.

Some transistors are manufactured to narrow tolerances, and others have fairly wide tolerances. A transistor curve tracer is very useful to select pairs of closely matched transistors for replacement in critical circuits. Note that although some service-type curve tracers utilize design short-cuts and trade-offs, a pair of closely matched transistors will produce

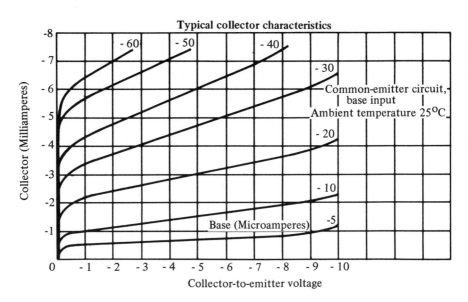

Fig. 7–12 Example of a collector family of characteristics specified for a small germanium transistor (Courtesy of RCA)

practically the same CRT patterns. Of course, an economy-type instrument might not be sufficiently accurate to evaluate a pattern precisely with respect to the collector family specified in a transistor manual. We will find that a few transistor curve tracers are designed for in-circuit application as well as out-of-circuit application. The same general limitations noted previously for the DC-beta type of tester apply also to in-circuit tests with a curve tracer.

7.5   FIELD-EFFECT TRANSISTORS

Field-effect transistors can be tested with the more elaborate service-type transistor testers, such as illustrated in Fig. 7–13. Conventional junction-type or bipolar transistors operate on the basis of electron and hole conduction. On the other hand, field-effect transistors (FET) are unipolar devices that operate on the basis of electron conduction, or of hole conduction, but not both. Whereas a bipolar transistor has emitter, base, and collector electrodes, an FET has source, gate, and drain electrodes. Operation of an N-channel FET is illustrated in Fig. 7–14. It is called an N channel device because the carriers flow in N-type semiconductor substance. The resistance of the channel is approximately 1000 ohms, measured from source to drain, or from drain to source.

Note that the gate in Fig. 7–14(b) is the control electrode. It may be a reverse-biased P-type electrode with a junction to the N channel. Such a device is called a junction field-effect transistor (JFET). Or, the gate may have a thin film of insulating substance between the P-type electrode and the N-channel substance. This is called an insulated-gate field-effect transistor (IGFET). When the gate is connected to the source, as in Fig. 7–14(b), current flows between source and drain as if the gate were absent. Next, when a negative bias is applied to the gate in (c), electrons in the N channel are repelled, and less current flows from source to drain. For example, the resistance of the N channel might be increased to 5000 ohms. Application of still more negative bias voltage will cut off or "pinch off" the electron flow, as depicted in (d).

Other FET's employ P-type substance for the channel and conduct by means of holes. The polarity of the supply voltages is reversed for P-channel devices. There is no gate-current flow in either type under normal conditions. Of course, if the gate is forward-biased in a JFET, gate current will flow. However, no gate current can flow in an IGFET unless the gate insulation breaks down. Breakdown can easily occur, as explained in greater detail subsequently. Many IGFET's have metallic-oxide insulation between the gate electrode and the channel substance. These are called metallic-oxide substrate field-effect transistors (MOSFET.) Figure

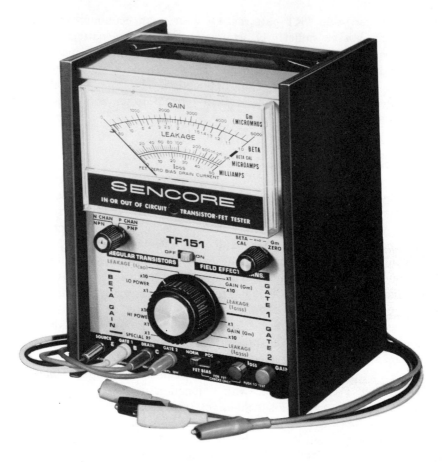

Fig. 7–13 Transistor tester for bipolar and field-effect devices
(Courtesy of Sencore)

7–14 shows an FET operating in the depletion mode. However, other FET's are designed to operate in the enhancement mode. This means that the gate polarity is reversed, and channel current is caused to flow by attraction, instead of by repulsion.

All JFET's operate in the depletion mode. Any depletion device normally exhibits current flow at zero bias. On the other hand, an enhancement device normally has practically no current flow until the gate is driven. The input impedance of an FET is typically 100 megohms or more. A JFET is normally operated in class A; this means that the gate is always reverse biased, so that gate current does not flow. However, a MOSFET may be operated in class B; this means that the gate is driven both positive and negative. An enhancement-mode MOSFET is usually operated in class C; this means that the gate is reverse biased, so that drain current flows only on drive peaks. Some MOSFET's have dual gates. In other MOSFET designs, a zener diode is built-in between the gate and source. This is a protective feature, to prevent accidental puncture of the insulation between the gate and the channel substance. Symbols for various types of FET's are shown in Fig. 7–15.

FET's are checked for gain in much the same manner as bipolar

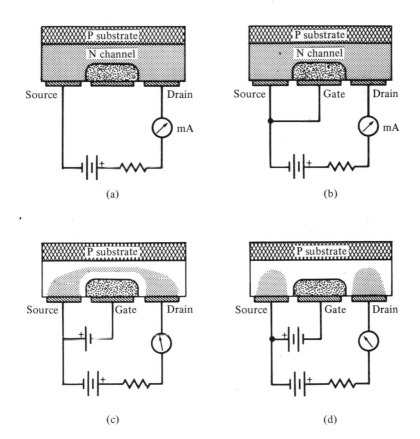

(a) (b)

(c) (d)

Fig. 7–14   Operation of an N-channel FET: (a) current (electrons) flows through the N channel as through an ordinary resistor; (b) the gate is the control electrode. (c) If the gate is negatively biased with respect to the source, current flow is decreased. (d) High negative bias on the gate causes "pinch off" and stops electron flow through the N channel.

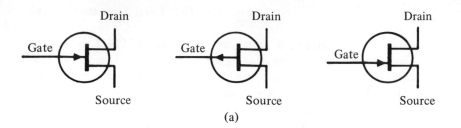

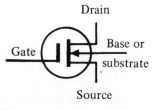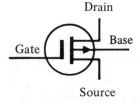

(b)

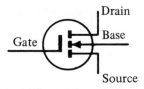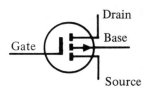

(c)

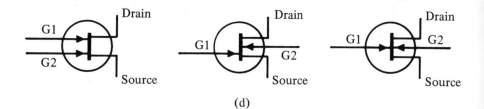

(d)

Fig. 7–15 Symbols for various types of FET's: (a) JFET's: Arrow points to N-type substance, away from P-type substance. (b) Depletion-type MOSFET's: Arrow points to N-type substrate, away from P-type substrate. (c) Enhancement-type MOSFET's: Arrow points to N-type substrate, away from P-type substrate. (d) Dual-gate FET's: N-channel symmetrical type is at left; N-channel nonsymmetrical type is at center; alternate symbol is at right.

transistors. The only difference is that since an FET is a voltage-operated device, the change in drain current is indicated for a given change in gate voltage, with test results in transconductance units. Leakage is checked by measurement of current flow (if any) between gate and source electrodes. This is called an I_{GSS} test, and denotes that the drain is tied to the source during the leakage test. An I_{DSS} test [Fig. 7–14(b)] is provided by some FET testers for the purpose of matching FET's in critical applications. As noted previously, MOSFET's are much more susceptible to accidental damage than bipolar transistors. For example, static charges that would be disregarded in handling bipolar transistors can ruin a MOSFET. Therefore, it is necessary to always keep the gate and source leads short-circuited before a MOSFET is connected into a circuit. Similarly, the gate and source leads must be short-circuited before a MOSFET is disconnected from a circuit. This precaution is not required in the case of MOSFET's with built-in zener diodes.

QUESTIONS AND PROBLEMS

True-False

1. In-circuit transistor tests are limited.
2. Diodes are checked for beta.
3. It is not possible to check a transistor for I_{CEO} in-circuit.
4. Measurement of the parameters of a transistor are the same for any values of bias voltage.
5. Beta is the ratio of collector current to base current.
6. The DC beta of a transistor is the same value as the AC beta.
7. The beta of a transistor is the same for any value of collector current.
8. The leakage current I_{CBO} is measured with the emitter open.
9. In-circuit beta tests are usually quite accurate.
10. If the collector and emitter leads are interchanged, the beta will read normally.
11. Leakage tests must always be made with the transistor out of the circuit.
12. The transistor curve tracer displays a family of curves for the transistor under test.
13. The curve tracer is an excellent in-circuit tester.
14. A transistor curve tracer is useful in selecting matched pairs.
15. Field effect transistors have a gate, a source, and an emitter.

Multiple-Choice

1. In-circuit transistor tests are
   (a) limited to a few checks.
   (b) more useful than out-of-circuit tests.
   (c) the only true test of a transistor's performance.

2. Most service-type transistor testers are designed to check
   (a) beta.
   (b) beta and leakage current.
   (c) beta, leakage current, and cut-off frequency.

3. In-circuit transistor tests are often useful as
   (a) a go-no-go form of quick check.
   (b) the only true test of a transistor's amplifying ability.
   (c) the only true leakage test.

4. The ratio of DC collector current to DC base current gives the
   (a) transistor beta.
   (b) transistor DC beta.
   (c) transistor AC or DC beta.

5. The leakage current I_{CEO} of a transistor is measured with the _____ open.
   (a) collector
   (b) base
   (c) emitter

6. The ratio of the leakage current I_{CEO} and I_{CBO} for a good transistor is approximately
   (a) one.
   (b) beta.
   (c) infinite.

7. Accuracy of an in-circuit beta test
   (a) depends on the components shunting the transistor.
   (b) is quite good.
   (c) depends on the type of transistor.

8. Testing an NPN transistor with the transistor tester in the PNP position will result in
   (a) a normal beta reading, but a high I_{CEO} reading.
   (b) a low I_{CEO} reading, but a normal beta reading.
   (c) no reading.

9. The in-circuit gain test becomes impractical when the
   (a) circuit shunting resistors are low.
   (b) circuit shunting resistors are high.
   (c) transistor is an NPN type.

10. The transistor curve tracer displays a family of transistor curves that can be used to determine
    (a) beta.
    (b) beta and I_{CEO}.
    (c) beta, I_{CEO}, and nonlinearity.

11. The bias between the gate and source on all JFET's is
    (a) reverse.
    (b) forward.
    (c) zero.

General

1. What are the common parameters that are measured with a transistor tester?
2. How is leakage generally measured in a transistor?

3. What is the purpose of most in-circuit transistor tests?
4. How is DC beta measured by the transistor tester in Fig. 7–3?
5. What are the limits to the beta measurements with an in-circuit transistor tester?
6. How can you make an out-of-circuit transistor test to a transistor that is mounted in a printed circuit board, without disconnecting the transistor?
7. What is the purpose of the transistor curve tester?
8. How is the base current varied in a transistor by a curve tracer?
9. How could you obtain a permanent record of a transistor characteristic from a curve tracer?
10. What are some of the problems that are encountered when testing FET's?
11. In a junction transistor the variable on the curve tracer was base current. What is the variable for an FET?

8

Sweep
and Marker
Generators

8.1 SWEEP-GENERATOR REQUIREMENTS

Television receivers utilize tuned circuits that have greater bandwidth than the tuned circuits in AM and FM broadcast receivers, or in communications receivers. For example, Fig. 8–1 shows a television signal spectrum for Channel 10. Since a total bandwidth of 6 MHz is involved in alignment procedures, considerable time would be required to check the frequency response of the tuned circuits with a conventional signal generator, using a point-by-point test procedure. Therefore, generators that automatically plot frequency-response curves on the screen of an oscilloscope are utilized in TV servicing procedures. This type of generator is called a sweep-frequency (or sweep) generator. It is essentially a frequency-modulated (FM) generator with sufficient deviation to sweep through a complete TV signal channel. We observe in Fig. 8–1 that if a test signal is varied from 47.25 MHz to 39.75 MHz, the receiver tuned circuits will respond with an output voltage variation that defines their over-all frequency response curve. This tuned-circuit output voltage is applied to the vertical input terminals of an oscilloscope in order to display the response curve on the CRT screen.

A sweep generator provides an FM test signal as depicted in Fig. 8–2. This signal has a center frequency, which is deviated back and forth at a 60-Hz rate. The amount of deviation may be as great as ±5 MHz in some TV alignment procedures. Note in Fig. 8–2 that the modulating voltage has a 60-Hz sine waveform. Although some other modulating frequency could be employed, it is convenient and economical to make use of the

169

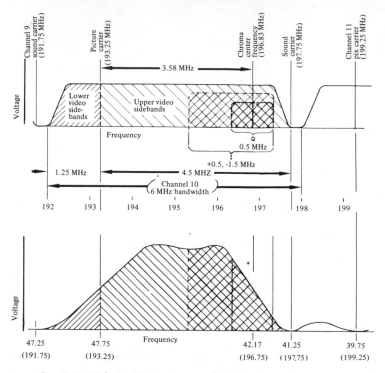

Fig. 8–1 Television signal spectrum for Channel 10, and typical frequency response of receiver-tuned circuits up to the video detector (Courtesy of B & K Mfg. Co.)

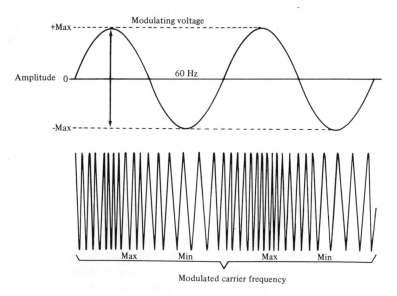

Fig. 8–2 Frequency-modulated test signal produced by a sweep generator

Channel No.	Freq (MHz) boundaries		Channel No.	Freq (MHz) boundaries		Channel No.	Freq (MHz) boundaries
	54		26	P 543.25 / S 547.75	542 / 548	55	P 717.25 / S 721.75	716 / 722
2	P 55.25 / S 59.75	60	27	P 549.25 / S 553.75	554	56	P 723.25 / S 727.75	728
3	P 61.25 / S 65.75	66	28	P 555.25 / S 559.75	560	57	P 729.25 / S 733.75	734
4	P 67.25 / S 71.75	72	29	P 561.25 / S 565.75	566	58	P 735.25 / S 739.75	740
5	P 77.25 / S 81.75	76 / 82	30	P 567.25 / S 571.75	572	59	P 741.25 / S 745.75	746
6	P 83.25 / S 87.75	88	31	P 573.25 / S 577.75	578	60	P 747.25 / S 751.75	752
7	P 175.25 / S 179.75	174 / 180	32	P 579.25 / S 583.75	584	61	P 753.25 / S 757.75	758
8	P 181.25 / S 185.75	186	33	P 585.25 / S 589.75	590	62	P 759.25 / S 763.75	764
9	P 187.25 / S 191.75	192	34	P 591.25 / S 595.75	596	63	P 765.25 / S 769.75	770
10	P 193.25 / S 197.75	198	35	P 597.25 / S 601.75	602	64	P 771.25 / S 775.75	776
11	P 199.25 / S 203.75	204	36	P 603.25 / S 607.75	608	65	P 777.25 / S 781.75	782
12	P 205.25 / S 209.75	210	37	P 609.25 / S 613.75	614	66	P 783.25 / S 787.75	788
13	P 211.25 / S 215.75	216	38	P 615.25 / S 619.75	620	67	P 789.25 / S 793.75	794
14	P 471.25 / S 475.75	470 / 476	39	P 621.25 / S 625.75	626	68	P 795.25 / S 799.75	800
15	P 477.25 / S 481.75	482	40	P 627.25 / S 631.75	632	69	P 801.25 / S 805.75	806
16	P 483.25 / S 487.75	488	41	P 633.25 / S 637.75	638	70	P 807.25 / S 811.75	812
17	P 489.25 / S 493.75	494	42	P 639.25 / S 643.75	644	71	P 813.25 / S 817.75	818
18	P 495.25 / S 499.75	500	43	P 645.25 / S 649.75	650	72	P 819.25 / S 823.75	824
19	P 501.25 / S 505.75	506	44	P 651.25 / S 655.75	656	73	P 825.25 / S 829.75	830
20	P 507.25 / S 511.75	512	45	P 657.25 / S 661.75	662	74	P 831.25 / S 835.75	836
21	P 513.25 / S 517.75	518	46	P 663.25 / S 667.75	668	75	P 837.25 / S 841.75	842
22	P 519.25 / S 523.75	524	47	P 669.25 / S 673.75	674	76	P 843.25 / S 847.75	848
23	P 525.25 / S 529.75	530	48	P 675.25 / S 679.75	680	77	P 849.25 / S 853.75	854
24	P 531.25 / S 535.75	536	49	P 681.25 / S 685.75	686	78	P 855.25 / S 859.75	860
25	P 537.25 / S 541.75	542	50	P 687.25 / S 691.75	692	79	P 861.25 / S 865.75	866
			51	P 693.25 / S 697.75	698	80	P 867.25 / S 871.75	872
			52	P 699.25 / S 703.75	704	81	P 873.25 / S 877.75	878
			53	P 705.25 / S 709.75	710	82	P 879.25 / S 883.75	884
			54	P 711.25 / S 715.75	716	83	P 885.25 / S 889.75	890

Fig. 8–3   Television picture- and sound-carrier frequencies

power-line frequency. Similarly, although a sawtooth modulating voltage could be utilized, it is convenient and economical to use the power-line waveform. Next, we will recognize that the carrier frequency which is frequency modulated must be set to a number of center-frequency values to accommodate the various tuned-circuit sections in a television receiver. For example, the ratio detector in the sound section of a TV receiver has a center frequency of 4.5 MHz. On the other hand, the UHF tuner in a TV receiver may operate at a center frequency as high as 887.5 MHz, as seen in Fig. 8–3.

8.2   SWEEP-GENERATOR FACILITIES
AND OUTPUTS

Figure 8–4 shows the panel layout of a typical VHF, IF, and video-frequency sweep generator. This instrument has a center-frequency range from approximately 4 kHz to 240 MHz. Continuous frequency coverage is provided, whereas similar generators may provide a switch control for the VHF range. Other generators may omit the FM range from 88 to 108 MHz. Still other generators may omit the video-frequency range from 40 kHz to 4.5 MHz. On the other hand, the more elaborate sweep generators provide UHF coverage from 470 to 890 MHz. Some sweep generators contain built-in frequency-marker facilities, and others must be supplemented by marker generators. In the example of Fig. 8–4, a front-panel socket is provided into which chosen marker crystals can be plugged. However, built-in continuous marker-frequency coverage is not included. Marking procedures are explained in greater detail subsequently.

We observe in Fig. 8–4 that four frequency ranges are provided. Two attenuator controls for the sweep-signal output permit level variation from approximately 0.1 volt maximum to 100 μV minimum. The anttenuator for the marker signal permits a wide variation in marker amplitude to be selected by the operator. Note that the sweep-width control varies the deviation of the FM test signal. A total sweep width of 12 MHz is available in this example. This is a deviation of ±6 MHz. Next, let us consider the function and operation of the phase control. As noted on the diagram, this control changes the phase of the horizontal-deflection voltage to the scope. The test arrangement is depicted in Fig. 8–5. Note that the horizontal amplifier of the scope is energized by a 60-Hz sine-wave voltage. The phase of this horizontal-deflection voltage must coincide with that of the 60-Hz modulating voltage in Fig. 8–2. Otherwise, the forward and return traces do not coincide in the screen pattern, as shown in Fig. 8–6. Therefore, the generator phase control is adjusted to bring the forward and return traces into "layover."

We observe in Fig. 8–6(c) that there is a residual double-image effect, even when the phase control is adjusted correctly. Therefore, most sweep generators have built-in return-trace blanking action. That is, the sweep oscillator is automatically switched off during the return-trace time, so that there is no output from the receiver under test during this time. The result of this blanking action is to display the response curve on a zero-volt reference line, as illustrated in Fig. 8–7. Of course, correct phase relations must also be maintained when the blanking facility is used; otherwise, the curve and baseline will be relatively displaced, and more or less of the curve will be "chopped off" at one end.

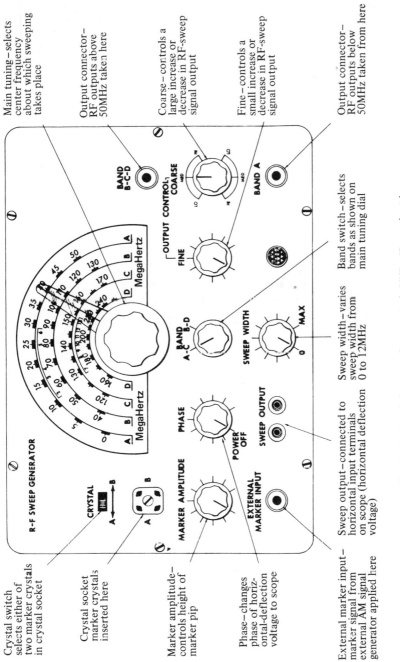

Fig. 8–4 Panel layout of a typical VHF, IF, and video-frequency sweep generator (Courtesy of Heath Co.)

Main tuning – selects center frequency about which sweeping takes place

Output connector – RF outputs above 50MHz taken here

Coarse – controls a large increase or decrease in RF-sweep signal output

Fine – controls a small increase or decrease in RF-sweep signal output

Output connector – RF outputs below 50MHz taken from here

Band switch – selects bands as shown on main tuning dial

Sweep width – varies sweep width from 0 to 12MHz

Sweep output – connected to horizontal input terminals on scope (horizontal deflection voltage)

External marker input – marker signal from external AM signal generator applied here

Phase – changes phase of horizontal-deflection voltage to scope

Marker amplitude – controls height of marker pip

Crystal socket – marker crystals inserted here

Crystal switch selects either of two marker crystals in crystal socket

R-F SWEEP GENERATOR

CRYSTAL
A ← → B

MegaHertz

MegaHertz

BAND B-C-D

OUTPUT CONTROL
COARSE
HI MED LO

FINE

BAND A

BAND A-C B-D

SWEEP WIDTH
0 MAX

PHASE

POWER OFF

SWEEP OUTPUT

MARKER AMPLITUDE

EXTERNAL MARKER INPUT

173

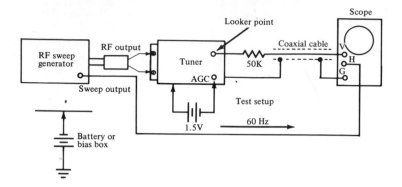

Fig. 8–5 Test arrangement for sweep-frequency alignment of a VHF tuner

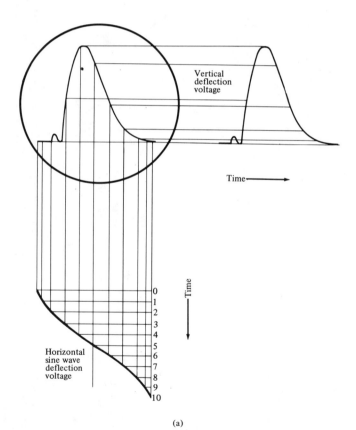

(a)

Fig. 8–6 Display of a response curve on a 60-Hz sine-wave time base: (a) development of CRT display;

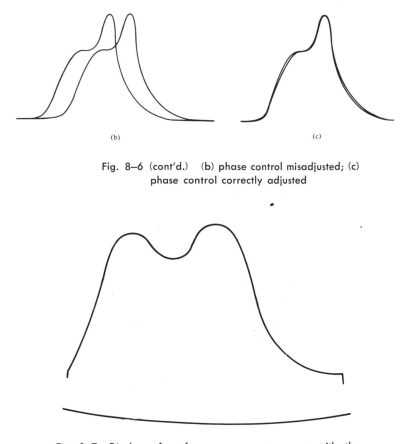

(b)

(c)

Fig. 8–6 (cont'd.) (b) phase control misadjusted; (c) phase control correctly adjusted

Fig. 8–7 Display of a frequency-response curve with the return trace blanked

8.3 MARKING A FREQUENCY-RESPONSE CURVE

Two principal methods are utilized to indicate specific frequencies along a frequency-response curve. These are termed the beat-marking method and the absorption-marker method. In Fig. 8–8 we observe that the frequency-sweeping process involves a frequency progression, shown in simplified form as $f1$, $f2$, and $f3$. It was evident that if we mix a CW signal with the sweep-frequency signal, that a beating process must occur, with zero beat occurring at the point where the sweep frequency is instantaneously equal to that of the CW marking signal. For example, if the

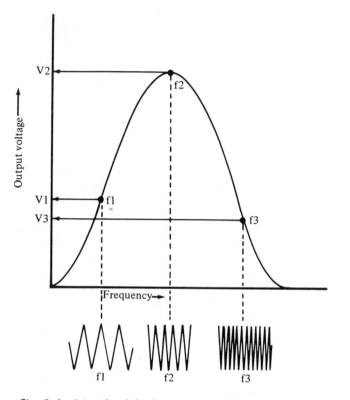

Fig. 8–8 Principle of the frequency-sweeping process

marking signal has a frequency equal to $f2$, zero beat will occur at the top of the response curve in Fig. 8–8. Note that the beat frequency, on the other hand, will be quite high at points $f1$ and $f3$. For example, we will suppose that $f2$ is equal to 43 MHz, that $f1$ is equal to 41 MHz, and that $f3$ is equal to 45 MHz. Then, with a beat frequency of zero at $f2$, we will have a beat frequency of 2 MHz at $f1$, and also at $f3$.

Next, with reference to Fig. 8–5, we observe that the test arrangement includes a 50K resistor in series with the scope input lead. This resistor is ordinarily called an isolating resistor; its function is to form a low-pass filter in combination with the capacitance of the coaxial input cable. It is evident that when the beat frequency produced by a marker signal passes through the low-pass RC filter, only the portion of the beat waveform in the vicinity of zero frequency will be passed in significant amplitude. The result, insofar as the CRT screen display is concerned, is seen in Fig. 8–9(a). We observe that the low-frequency beat interval appears as a "pip" or "birdie" on the response curve. Of course, unless low-pass filtering action is provided, the beat marker would appear merely as a "fuzz" or thickening of the response-curve trace.

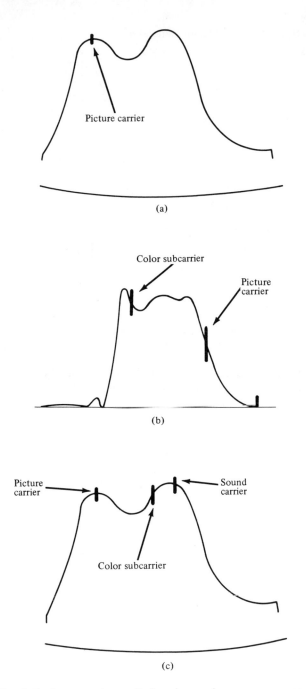

Fig. 8–9 Beat markers displayed on frequency-response curves: (a) picture-carrier marker on a VHF response curve; (b) picture-carrier and color-subcarrier markers on an IF response curve; (c) picture-carrier, sound-carrier, and color subcarrier markers on a VHF response curve

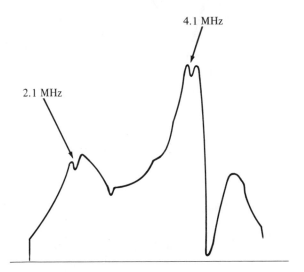

Fig. 8–10   Appearance of absorption markers on a response curve

It is logical to ask how the response curve can be displayed without distortion when a low-pass filter is employed. The answer is that if the filter has a time-constant that is sufficiently short to pass a 60-Hz square wave without objectionable distortion, it will pass the response-curve signal without objectionable distortion. For example, if we utilized a 250K isolating resistor, instead of a 50K resistor, the displayed response curve would be substantially distorted. Next, we will observe that some sweep generators have multiple-marking facilities. For example, Fig. 8–9(b) shows dual markers at the picture-carrier and color-subcarrier frequency points on an IF response curve. Again, Fig. 8–9(c) shows triple markers at the picture-carrier, sound-carrier, and color-subcarrier frequency points on a VHF response curve. In case a sweep generator does not have built-in beat-marking facilities, we simply use a separate marker generator and mix the CW output signal from the marker generator with the sweep output signal from the sweep generator.

Although beat markers are in widest general use, they are unsuitable for certain applications such as video-frequency response-curve marking. Therefore, another type of marker indication is utilized, called absorption markers. The appearance of absorption markers on a chroma bandpass-amplifier response curve is illustrated in Fig. 8–10. An absorption marker places a small "dip" in the trace. Beat markers are generally unsuitable in video-frequency sweep applications because the video-frequency sweep output from a generator tends to contain residual spurious frequencies, as explained in greater detail subsequently. In turn, a beat-marker signal will heterodyne with these residuals, and may form confusing spurious markers on a curve. On the other hand, an absorption marker employs a simple trap action and has no residual spurious frequencies. Some sweep generators have built-in absorption-marking facilities; others are supplemented by absorption-marker boxes. The marker box is connected in series with the video-frequency sweep-output cable.

Note in Fig. 8–11(a) that each trap coil is connected to an external metal disk. These are test terminals which are utilized to identify a particular absorption marker on a response curve. As seen in Fig. 8–10, four absorption markers are displayed on the curve. If we need to identify the 4.1-MHz marker, for example, we will touch a finger to the "4.1" terminal on the marker box. In turn, body capacitance will detune the trap circuit,

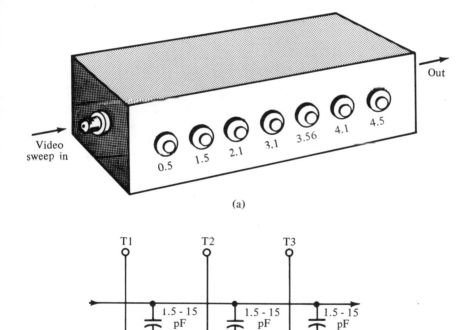

(a)

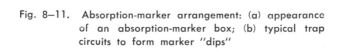

(b)

Fig. 8–11.   Absorption-marker arrangement: (a) appearance of an absorption-marker box; (b) typical trap circuits to form marker "dips"

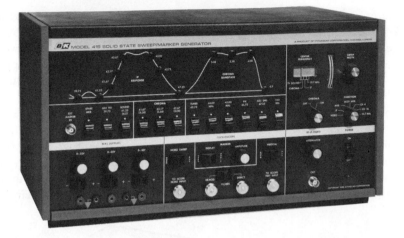

Fig. 8–12 Appearance of a sweep-frequency generator designed for color-TV service (Courtesy of B & K Mfg. Co.)

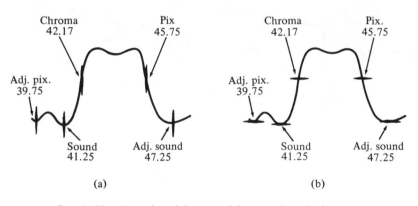

Fig. 8–13 Vertical and horizontal beat-marker displays: (a) vertical presentation; (b) horizontal presentation

and the marker will move somewhat toward the low-frequency end of the curve. This movement identifies the particular frequency marker of interest.

Some sweep-frequency generators are designed for maximum utility in color-TV servicing applications, and have indicator lamps that glow on typical response curves engraved on the front panel of the instrument, as illustrated in Fig. 8–12. For example, if a 3.58-MHz marker signal is switched into the output system, the lamp marked "3.58" will glow on the outline of the chroma response curve. Thereby, the operator is not only reminded of the marker frequency that is in use, but also of its proper location on the response curve. In addition, the instrument shown in Fig. 8–12 provides a choice of vertical or horizontal beat-marker indication. The distinction between these two forms of marker display is exemplified in Fig. 8–13. Vertical markers appear to best advantage along the top of a curve, but might be almost invisible on steep sides of the curve. On the other hand, horizontal markers appear to best advantage on steep sides of a curve, but might be almost invisible along the top of the curve.

8.4 SWEEP-SIGNAL GENERATION

Various types of FM oscillators are utilized in sweep-frequency generators. Of these, the increductor arrangement depicted in Fig. 8–14 has become comparatively popular. An increductor is a coil wound on a ferrite core, with a 60-Hz control winding. The 60-Hz winding produces a 60-Hz magnetic field that varies the permeability of the ferrite core. In turn, the inductance of the coil varies at a 60-Hz rate. This coil is an oscillator tank circuit, so that the oscillator frequency deviates back and forth at a 60-Hz rate. To obtain maximum linearity of deviation, a DC bias current is passed through the control winding, in addition to the 60-Hz AC current. Four frequency bands are provided in the example of Fig. 8–14.

It is evident that a swept oscillator should supply uniform output throughout its interval of deviation. That is, if the sweep-frequency output is not "flat," the resulting response curves displayed on the scope screen will be tilted and distorted in shape. Therefore, the FET in Fig. 8–14 is AGC-controlled, so that the sweep-frequency output is maintained practically uniform over its interval of deviation. The lossy choke also provides an impedance characteristic that assists in maintaining uniform signal amplitude. Note that a blanking voltage is applied to the gate of the FET, to switch the oscillator off during the return-trace interval. This is a 60-Hz square-wave voltage. Thereby, a single-trace display is obtained, with a zero-volt reference line on the scope screen. A zero-volt reference line is useful in wide-band alignment procedures because it may be impossible to completely sweep the curve. In turn, the ends of the curve are "cut off," as seen in Fig. 8–7. However, the zero-volt reference line will always indicate the base level of the curve.

Most sweep-frequency generators provide push-pull (double-ended) output on the VHF range to match the double-ended input of TV tuners. Single-ended output is provided on IF, video-frequency, and chroma bands. It is easy to check the flatness of the sweep-signal output on the VHF range with a double-ended demodulator probe, as depicted in Fig. 8–15(a). Normally, a practically flat swept trace is observed on the scope screen. Single-ended output signal voltages from the sweep generator can

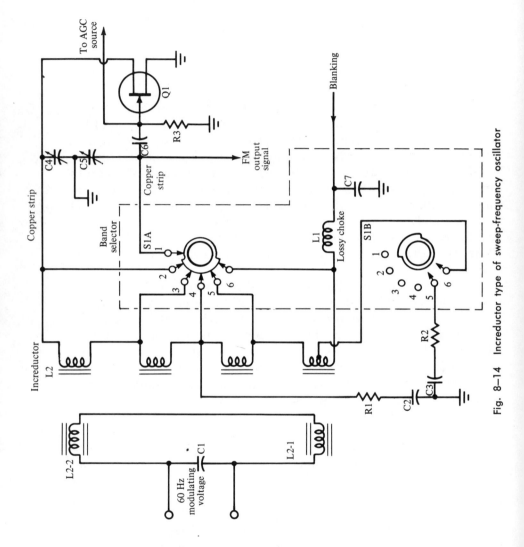

Fig. 8–14 Increductor type of sweep-frequency oscillator

be checked similarly by using the single-ended demodulator probe shown in Fig. 8–15(b). If a sweep generator has single-ended output on its VHF function, the single-ended signal can be changed into a double-ended signal for all practical purposes by means of a resistor pad, as depicted in Fig. 8–16. A single-ended output facility generally employs a 75-ohm cable. Therefore, we choose a pad arrangement that converts the 75-ohm output to a 300-ohm output to match the input impedance of a VHF tuner.

Service-type sweep-frequency generators do not include individual UHF sweep oscillators. However, harmonics of the VHF sweep signal may be adequate to check UHF tuners. Since the harmonic output voltage is quite low, a UHF tuner must be checked by connecting the scope at the

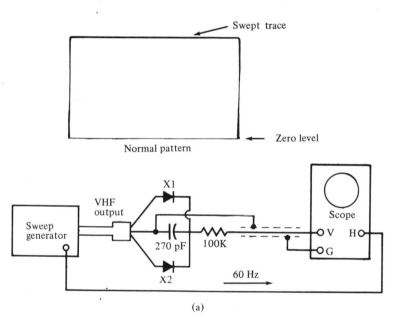

(a)

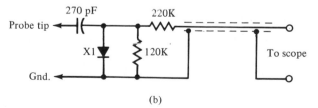

(b)

Fig. 8–15   Demodulator probe configurations: (a) double-ended probe for VHF check of sweep-generator output; (b) single-ended demodulator probe used in general service work

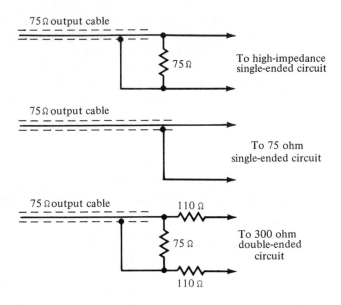

Fig. 8–16 Output-cable terminations for different applications

output of the picture detector in the receiver. Thereby, adequate gain is obtained for a normal pattern display, and alignment of the UHF tuner is evaluated on the basis of the over-all response curve. That is, the over-all response curve should have essentially the same characteristics as the IF response curve because a UHF tuner normally has a comparatively wide pass band.

Video-frequency sweep signals are generated by the beat method, as depicted in Fig. 8–17. This is the only practical method of developing an output frequency range from 40 kHz to 4.5 MHz by reasonably uncomplicated design principles. A fixed RF oscillator and a variable RF oscillator operate as the signal sources. The fixed oscillator might supply a 100-MHz CW signal, while the variable oscillator might supply a 102.5-MHz signal that varies (deviates) between the limits of 100 MHz and 105 MHz. We will consider an instant at which the beat frequencies are 100 MHz and 102.5 MHz. In turn, the beat waveform has an envelope frequency of 2.5 MHz. The heterodyne detector eliminates the negative half-cycles of the beat waveform, leaving a pulsating-DC voltage that has an average value with a frequency of 2.5 MHz. This average-value waveform is separated from the RF component by means of a low-pass filter. In turn, the video-frequency sine waveform with a frequency of 2.5 MHz is provided.

In practice, the video-frequency signal is not quite pure, because of feed-through by harmonics and cross-beat frequencies. That is, the RF oscillators do not have a perfect sine waveform, and the heterodyne detector does not have an ideal mixing characteristic. Moreover, no low-pass filter is perfect. However, the residual spurious outputs from a well-designed video-frequency sweep generator are quite low in level.

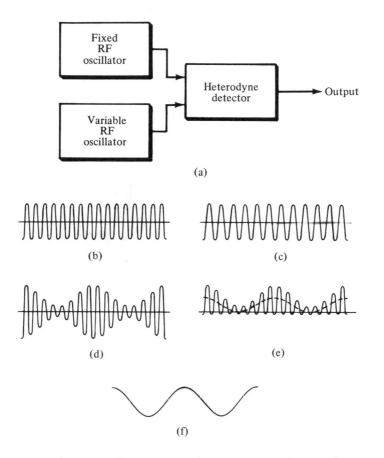

(a)

(b)   (c)

(d)   (e)

(f)

Fig. 8–17   Beat-frequency signal generation: (a) beat oscillators and heterodyne detector; (b) fixed RF input
signal; (c) variable RF input signal; (d) beat waveform; (e) heterodyne detector output; (f) filtered
difference-frequency output

8.5   VIDEO SWEEP MODULATION

Video sweep modulation (VSM) alignment techniques are used to check the over-all frequency response of color-TV receivers. This method is preferred because numerous tuned circuits are involved, and these

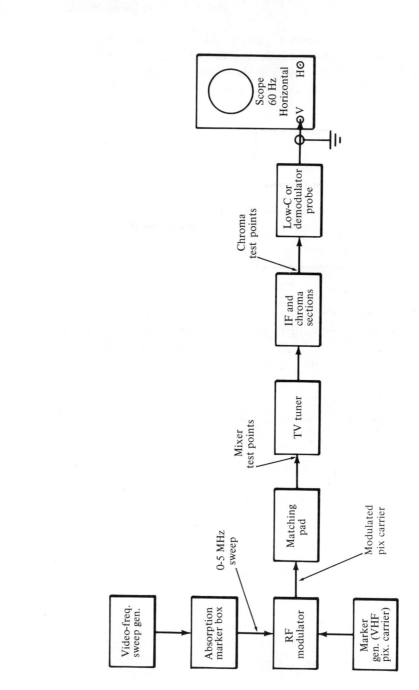

Fig. 8-18 Arrangement of a VSM alignment setup

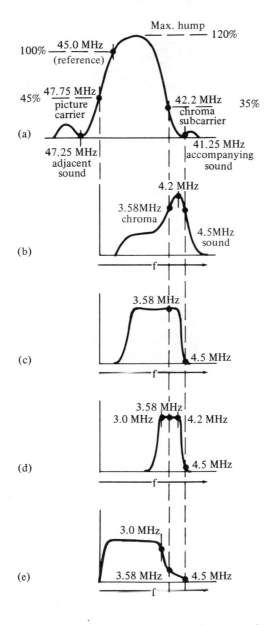

Fig. 8–19 Frequency response curves at the output of various sections in a typical color-TV receiver: (a) IF response curve; (b) bandpass-amplifier response curve; (c) overall bandpass and IF response curve; (d) chroma-demodulator response curve; (e) Y-amplifier response curve

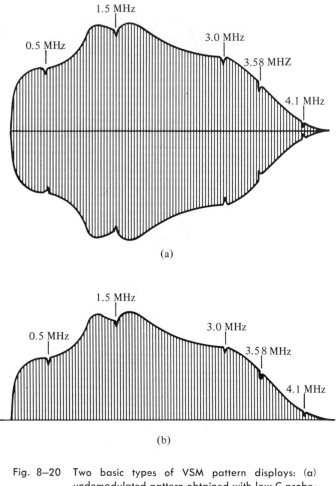

(a)

(b)

Fig. 8–20   Two basic types of VSM pattern displays: (a)
undemodulated pattern obtained with low-C probe;
(b) corresponding pattern obtained with demodulator probe

circuits must work together properly as a team. The best method of evaluating this teamwork is to make a VSM frequency-response check. A
typical alignment setup for a VSM check is shown in Fig. 8–18. This
arrangement employs a video-frequency sweep generator with an absorption-marker box. In turn, a semiconductor-diode modulator unit is energized by the video-frequency sweep signal and by the CW output from
a marker generator tuned to the picture-carrier frequency. Thereby, the
video-frequency sweep signal is said to be encoded into the CW carrier
signal. This VSM signal is then processed (in this example) through the
tuner, IF, and chroma sections of the color-TV receiver. The scope may
be applied via a low-capacitance probe or a demodulator probe.

Figure 8–19 shows frequency response curves that are normally observed at the output of various sections in a typical color-TV receiver. Thus, if the scope is applied at the output of a chroma demodulator in Fig. 8–18, a curve similar to that shown in Fig. 8–19(d) would be anticipated. Or, if the scope were applied at the Y-amplifier output, a curve as shown in (e) would be displayed in normal operation. Finally, we will observe that the type of CRT display depends on whether a demodulator probe or a low-capacitance probe is used with the scope. For example, Fig. 8–20 shows an undemodulated and a demodulated VSM pattern. These are demodulated displays; the undemodulated pattern includes the video-frequency sweep signal, whereas the demodulated pattern shows its envelope.

8.6   MARKER ADDERS

Most modern sweep-frequency generators contain built-in marker adders. This is a mixer arrangement that develops a constant-amplitude beat marker, which is equally visible on top of a response curve, or down in a trap on the base line. A marker adder operates by mixing the sweep and marker signals, and feeding the "pip" voltage directly to the scope, instead of passing the "pip" through the tuned circuits of the receiver under test. Thereby, the marker amplitude remains constant at all times and is not attenuated by trap action in the receiver circuits. This is a considerable convenience, not only in marking trap frequencies on response curves, but also in marking the center frequency of ratio-detector response curves. That is, a ratio detector rejects amplitude modulation almost completely, thereby making direct frequency-marking of S curves very difficult. However, with the aid of a marker adder, a beat marker is displayed at any desired amplitude on an S curve.

QUESTIONS AND PROBLEMS

True-False

1. The marker generator is used to indicate a particular frequency on the sweep pattern.
2. A sweep generator provides both an FM and an AM test signal.
3. The deviation frequency is the range of the sweep frequency.
4. A phase control minimizes the overlay effect.
5. The beat marker and the absorption marker methods are used to indicate specific frequencies along a frequency response curve.
6. The 50K resistor in Fig. 8–5 reduces the input capacitance.
7. Some sweep generators have multiple-marking facilities.
8. Beat markers are not suitable for video-frequency response-curve marking.
9. An increductor is a type of oscillator coil.

10. Most sweep-frequency generators provide single-ended output in the VHF range.
11. A single-ended signal can be changed into a double-ended signal by using a resistor pad.
12. The input impedance of a VHF tuner is 300 ohms.
13. Service-type sweep-frequency generators include individual UHF sweep oscillators.
14. Video-frequency sweep signals are generated by the beat method.
15. A marker adder is used to produce a beat marker on the S curve of the ratio detector.

Multiple-Choice

1. The purpose of using a marker generator with a sweep generator is to
   (a) extend the range of the sweep generator.
   (b) mark the frequency at any point on the response curve.
   (c) control the sweep frequency of the sweep generator.
2. The sweep frequency rate is usually
   (a) 60 Hz.
   (b) 120 Hz.
   (c) 39 MHz.
3. Adjustment of the phase control
   (a) prevents nonlinearity of the scope pattern.
   (b) prevents drifting of the FM signal.
   (c) adjusts the pattern for overlay.
4. The purpose of the 50K resistor in the test arrangement in Fig. 8–5 is to
   (a) reduce the effects of input capacitance.
   (b) act as a low-pass filter.
   (c) reduce the input signal.
5. If the beat marker appeared only as a thickening of the response-curve trace, the problem is probably
   (a) loss of beat signal.
   (b) not using a detector probe.
   (c) not using an isolating resistor.
6. An absorption marker places _____ on the trace.
   (a) a peak
   (b) fuzz
   (c) a dip
7. The frequency of an increductor coil is controlled by
   (a) a DC bias.
   (b) a 60-Hz AC signal.
   (c) a control oscillator.
8. The flatness of the sweep-signal output on the VHF range is tested with
   (a) an oscilloscope.
   (b) a TVM.
   (c) a double-ended demodulator probe.

9. A single-ended output generally employs a _____ ohm cable; therefore, we use a pad arrangement that converts the output to _____ ohms to match the input impedance of the VHF tuner.
   (a) 75; 75
   (b) 75; 300
   (c) 300; 75
10. Video-sweep modulation alignment techniques are used to check the overall frequency response of
    (a) the tuner.
    (b) FM sound section.
    (c) a color TV receiver.

General

1. Why is it necessary to use a sweep generator to align a television receiver?
2. What is a sweep generator?
3. What is the normal sweep frequency?
4. How is a specific frequency identified along a frequency-response curve?
5. How do we develop the marker signal when the sweep generator does not contain a marker generator?
6. What type of marker produces a small dip in the trace?
7. What is an increductor?
8. What is the purpose of the video sweep modulation alignment technique?
9. What is a marker adder?
10. What is the correct procedure for aligning a TV receiver?

9

FM
Stereo-Multiplex
Generators

9.1 INTRODUCTION

Stereophonic sound reproduction requires a pair of signals that char-
acterize the transmission and reception action and reaction that is associated with binaural hearing. Thus, the stereophonic process basically
involves a pair of microphones and a pair of speakers, as depicted in Fig.
Fig. 9–1, these two signals are generated by a pair of microphones, and the
the signal at the right ear. The individual audio signals in a stereophonic
system are termed "Left" (L) and "Right" (R) signals. In the example of
Fig. 9–1, these two signals are generated by a pair of microphones, and the
L signal differs more or less from the R signal in signal-component amplitudes and phases.

Note that the amplifier in Fig. 9–1 is actually a pair of amplifiers
called L and R amplifiers. Although energized by the same power supply,
each amplifier processes an individual signal. The output signals from the
amplifiers are fed to a pair of speakers. These speakers are placed an
appreciable distance apart in order to simulate the distance between the
microphones at the input end of the system. We will find that an FM-stereo
signal is basically the same as the simple audio system that has been
described. That is, an FM-stereo signal consists basically of two audio-
frequency signals corresponding to two pairs of sidebands that occupy
the same FM channel. An FM stereo-multiplex generator is a simplified
form of transmitter. Let us see how two signals can occupy the same
channel without interference.

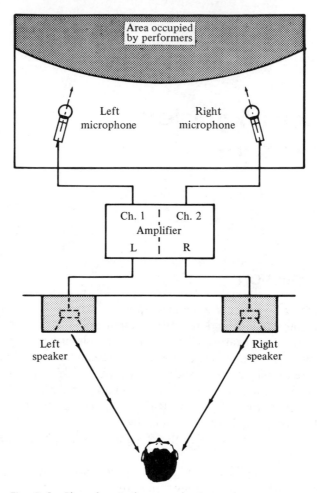

Fig. 9–1   Plan of a simple stereophonic system

9.2   PRINCIPLES OF STEREO MULTIPLEXING

One might surmise that an FM radio channel could be divided into two parts for transmission of L and R stereo signals on individual carriers; however, this is not a practical approach. A single audio signal occupies virtually all of the available bandwidth in an FM channel. In Fig. 9–2, we note that the maximum available deviation in each channel is ±75 kHz. A deviation of ±75 kHz corresponds to maximum modulation of the FM carrier. High-fidelity sound reproduction requires audio frequencies up to 15 kHz. Even if the transmitted audio frequencies were limited to 7.5 kHz, more than half of the FM channel would be occupied by the signal at

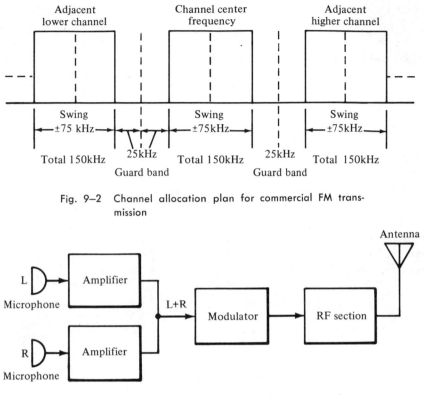

Fig. 9–2   Channel allocation plan for commercial FM transmission

Fig. 9–3   Addition of L and R signals produces a monophonic signal

maximum modulation. Accordingly, let us consider how a pair of signals can be effectively multiplexed in an FM channel.

When an appropriate multiplexing method is used, each audio signal occupies virtually all of the available bandwidth in the FM channel during maximum modulation. High-fidelity sound reproduction requires that both of the signals be transmitted without significant distortion, and without mutual interference. It is evident that each of the signals must have an electrical characteristic that permits practically complete separation from the combined signal in the reproducing process. In addition, FM stereo multiplexing has a compatibility requirement. That is, the multiplex signal must operate a conventional FM receiver normally, but must provide separable L and R signals to a stereo-multiplex receiver. This is accomplished in the approved FCC system.

We recognize that the addition of L and R audio signals, shown at left in Fig. 9–3, provides the equivalent of a conventional monophonic signal. That is, the electrical addition of the L and R signals in a mono signal can be modulated on an FM carrier with the same result as if a single microphone were employed. A conventional FM receiver responds to this signal as if it were an ordinary FM signal. Next, let us consider how additional information can be inserted independently in this type of system. If a 38-kHz carrier is introduced with the mono signal, Fig. 9–4(a), this carrier signal cannot be heard at the receiver because its frequency is above the audible range. The frequency relations in this situation are illustrated in Fig. 9–4(b).

Next, we will modulate the amplitude of this 38-kHz carrier (strictly speaking, it is a 38-kHz subcarrier) with an arbitrary audio signal S2, as seen in Fig. 9–5(a). Accordingly, sidebands are generated on each side of the subcarrier, as shown in Fig. 9–5(b). Note that if the audio signal S2 includes frequencies up to 15 kHz, these sidebands will include frequencies from 23 kHz to 53 kHz. The subcarrier and its sidebands are combined with the L + R mono signal, and then frequency-modulated on the FM carrier. This is the basic multiplexing, or encoding, process with which we are concerned. We recognize that if the signal in Fig. 9–5(b) is processed

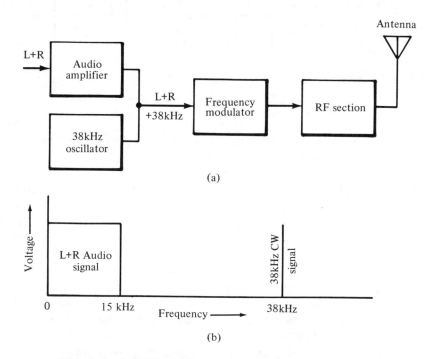

Fig. 9–4   A 38-kHz subcarrier is introduced with the L + R mono signal: (a) block diagram; (b) frequency spectrum

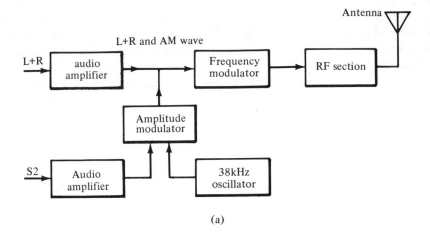

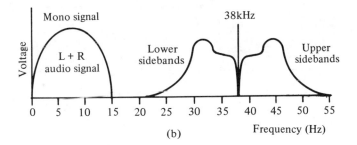

Fig. 9–5   Formation of a multiplex signal: (a) block diagram of the basic system; (b) frequency spectrum
of the modulating signal

by a conventional FM receiver, the 38-kHz subcarrier and its sidebands are
inaudible.

However, if a multiplex unit or adapter is employed with a conventional
FM receiver, both of the transmitted signals can be reproduced as shown
in Fig. 9–6. We observe that the upper speaker can reproduce only the
L + R signal which has a frequency range from nominally zero to 15 kHz.
Even if the upper speaker were capable of reproducing a signal in the
range from 23 kHz to 53 kHz, the human ear cannot respond to frequencies
in this range. Accordingly, the S2 signal is rejected by the upper speaker.
However, the S2 signal can be recovered by employing a 23- to 53-kHz
bandpass filter, and an AM detector to decode the multiplexed signal; the
lower speaker then reproduces the audio modulation envelope of the S2
signal.

Next, let us consider the form of the S2 signal that must be employed in order that the speakers in Fig. 9–6(a) will reproduce L and R signals. We know that an L + R signal is the equivalent of a mono signal. Similarly, at this point in our development of the system, the S2 signal is simply another independent mono signal. In Fig. 9–7 a stereo-multiplex system is arranged so that the right microphone produces an output from the right speaker and the left microphone produces an output from the left speaker.

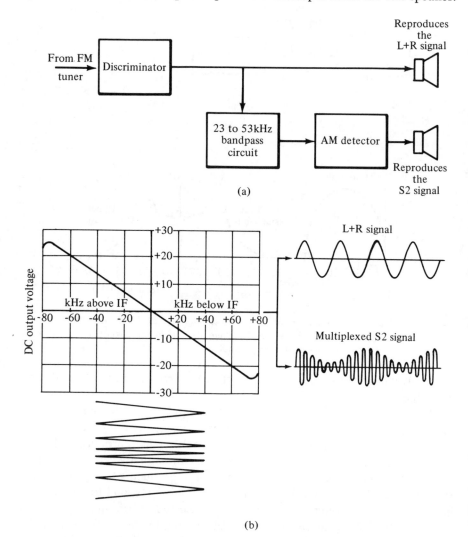

(a)

(b)

Fig. 9–6   Separation and reproduction of·multiplexed signals:
(a) arrangement for decoding the multiplex signal;
(b) discriminator processing of the signals

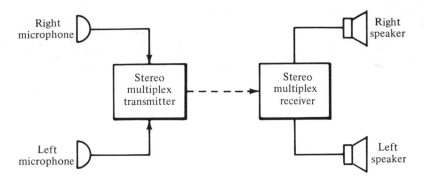

Fig. 9–7   Basic stereo-multiplex system

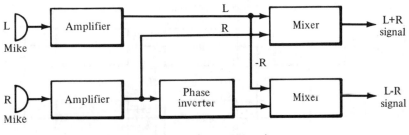

Fig. 9–8   Forming L + R and L − R signals

Note that the upper speaker in Fig. 9–6 would reproduce an R signal if it were energized by both an L + R signal and a −L + R signal. That is, the L components would cancel out, and a 2R signal would be fed to the upper speaker.

Next, we recognize that the lower speaker in Fig. 9–6 would reproduce an L signal if the speaker were energized by both an L + R and an L − R signal. The R components would cancel out in this case, and a 2L signal would be applied to the lower speaker. The first requirement is to form an L − R signal at the transmitter; this is accomplished as follows. With reference to Fig. 9–8, the upper mixer provides an L + R signal, as explained previously. Note that the output from the R microphone also energizes a phase inverter, which changes the R signal into a −R signal. Since the lower mixer processes both an L signal and a −R signal, it provides an L − R signal. Thus, the basic L + R and L − R signals employed by a stereo-multiplex system are generated for processing into the complete FM stereo-multiplex signal.

As shown in Fig. 9–9, the L − R signal is amplitude-modulated on the 38-kHz subcarrier. This modulated wave is combined with the L + R signal and applied to the FM modulator. In turn, the two signals are frequency-modulated on the RF carrier. The L + R signal "looks" like a

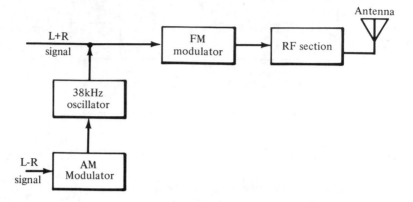

Fig. 9–9 Forming the composite signal

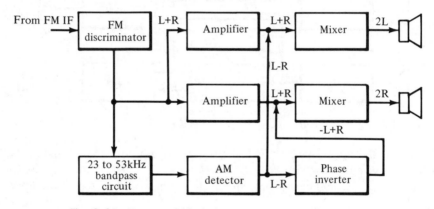

Fig. 9–10 Functional block diagram of the multiplex adapter

conventional mono signal, and the L − R signal is encoded into the composite FM stereo-multiplex signal. Next, let us consider the arrangement employed in the multiplex receiver. The discriminator output provides the frequencies depicted in Fig. 9–10. That is, the encoded L − R signal comprises frequencies from 23 to 53 kHz, and this band of frequencies is rejected by an audio amplifier. On the other hand, the encoded L − R signal passes through the 23- to 53-kHz bandpass circuit and is applied to the AM detector. In turn, the demodulated L − R signal from the AM detector is fed to the input of the upper mixer, thus combining the L + R signal with the L − R signal. Since the R components cancel, the upper mixer is driven by a 2L signal, and the upper speaker reproduces this L signal.

We observe in Fig. 9–10 that the output from the AM detector is also fed to a phase inverter. Accordingly, a −L + R signal is obtained, which is fed to the input of the lower mixer. Combination of the −L + R signal with L + R signal results in cancellation of the L component, and the lower mixer is driven by a 2R signal. Thus, the lower speaker reproduces this R signal. The bandpass section has the frequency characteristic illustrated in Fig. 9–11. This is the most basic arrangement for a multiplex adapter. However, we should note that other arrangements are commonly employed that dispense with filter action and operate on other signal-processing principles.

We must consider the signal modification employed by FM stereo-multiplex transmitters and in standard generators. This modification of the basic system entails suppression of the 38-kHz subcarrier by means of a balanced modulator, and insertion of a 19-kHz pilot subcarrier. The reason for this modification is that considerable energy is added to the FM signal when the L − R signal is encoded with the L + R signal. In order to obtain a good signal-to-noise ratio, the energy of the L − R signal is

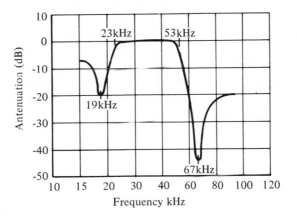

Fig. 9–11   Bandpass filter frequency-response curve

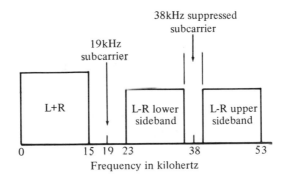

Fig. 9–12   Composite signal used to frequency-modulate the RF carrier

reduced by suppressing the 38-kHz subcarrier. Thus, only the upper and lower sidebands of the L − R signal are present in the composite signal. Since the 38-kHz subcarrier carries no information, it can be reinserted with the sidebands at the receiver. Let us see what is involved in this process.

It can be shown that the reinserted 38-kHz subcarrier must have the same phase, as well as the same frequency, as the subcarrier that was suppressed in the multiplexed signal. The only practical method of meeting these requirements is to provide a pilot for reference by the receiver system. We will perceive that this pilot signal cannot be a low-amplitude 38-kHz subcarrier because the upper and lower sidebands of the L − R signal flank the subcarrier very closely. In turn, it is impractical to filter this type of pilot signal from the composite signal at the receiver. Therefore, a sub-harmonic of the subcarrier is utilized; as seen in Fig. 9–12, the 19-kHz subharmonic falls in an unoccupied frequency interval between the L + R signal and the L − R lower sideband. Accordingly, it is a simple matter to pick out the 19-kHz pilot subcarrier from the composite signal at the receiver.

To reconstitute the 38-kHz subcarrier at the receiver, a local 19-kHz oscillator is locked with (synchronized by) the 19-kHz pilot subcarrier. The second harmonic in the output of the locked oscillator is picked out by a tuned circuit and mixed with the L − R upper and lower sideband signals from the discriminator. Since the phase of the reconstituted subcarrier is the same as in the original L − R signal before the subcarrier was sup-pressed, the original composite signal is made available for recovering the L and R signals. Although the foregoing multiplexing process is somewhat involved, it provides high-fidelity stereophonic sound reproduction in an FM channel that normally accommodates only a high-fidelity monophonic signal.

9.3 STEREO-MULTIPLEX WAVEFORMS

An FM stereo-multiplex generator (see Fig. 9–13) is a fixed-tone trans-mitter in miniature. It employs the basic principles that have been dis-cussed, and develops various waveforms as explained below. First, let us suppose that an R signal is being generated; this is typically a 1-kHz tone signal. The composite signal waveform appears as illustrated in Fig. 9–14. Let us analyze this waveform into its components. The R signal in this example is a 1-kHz sine wave. The L signal is not present (zero). With reference to Fig. 9–8, the outputs will consist of +R and −R signals (the −R signal is shifted 180° in phase with respect to the +R signal).

Next, with reference to Fig. 9–9, the −R signal is applied to an AM modulator, whereby it modulates the 38-kHz subcarrier. As explained previously, a balanced modulator is employed in this section in order to

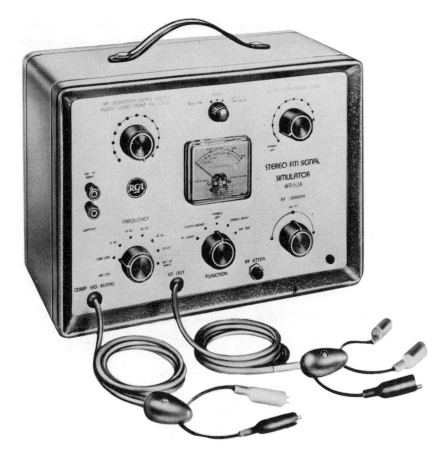

Fig. 9–13   Stereo FM signal generator (Courtesy of RCA)

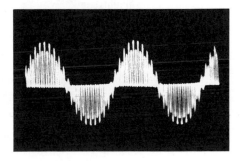

Fig. 9–14   Composite signal waveform (Courtesy of RCA)

suppress the subcarrier from the output waveform. Thus, the upper and lower −R sidebands are mixed with the +R signal and applied to the FM modulator. In terms of waveforms, this signal processing proceeds as shown in Fig. 9–15. That is, the complete L − R signal has a sinusoidal envelope, as shown in Fig. 9–15(a). Suppression of the carrier component results in the half-sine envelope of the modified L − R signal shown in Fig 9–15(b).

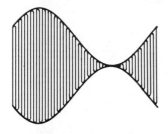

Subcarrier and both sidebands

(a)

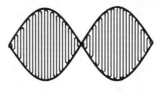

Sidebands without subcarrier

(b)

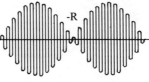

Sine-wave signal

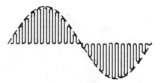

Sidebands without subcarrier

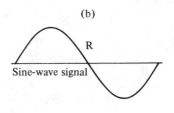

Sine-wave signal plus upper and lower sidebands

(c)

Fig. 9–15   Various waveforms in the multiplex generator circuitry: (a) complete L − R signal; (b) suppression of subcarrier results in a double-frequency modulation envelope; (c) formation of multiplex signal from addition of sine-wave signal to the sidebands with a suppressed subcarrier

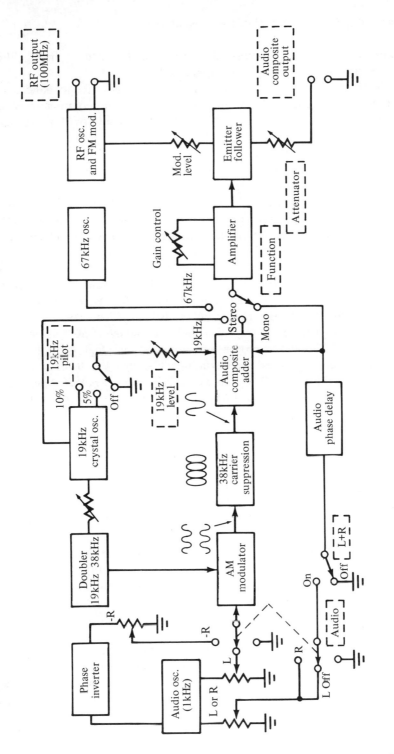

Fig. 9–16 Block diagram of a simple FM stereo-multiplex generator (Courtesy of Hickok Elec. Inst. Co.)

As noted previously, a 19-kHz pilot subcarrier is added to the signal after the 38-kHz subcarrier has been suppressed.

Next, the +R 1-kHz sine wave is added to the −R sidebands, as depicted in Fig. 9–15(c). The result is that the sideband component is superimposed (rides on) the 1-kHz sine-wave component. In this manner the waveform illustrated in Fig. 9–14 is generated. However, there is one additional component in the waveform of Fig. 9–14 that does not appear in Fig. 9–15(c). This is the 19-kHz pilot subcarrier, which appears as the protruding "spikes" in the waveform of Fig. 9–14. Note that the 19-kHz pilot subcarrier is added to the L + R and L − R signals in a composite adder, as seen in Fig. 9–16. The pilot subcarrier normally has an amplitude of 10%, but this can be reduced to 5%, if desired, to check receiver synchronization capability.

The L and R waveforms from the generator appear to be the same when displayed on a scope screen. This similarity results from the fact that a signal distinction can be made only in terms of phase difference. Note in Fig. 9–16 that a 67-kHz test signal is also provided. This signal is used for alignment of the SCA trap in a multiplex adapter. SCA is an abbreviation for the subsidiary carrier assignment signal that is transmitted by some FM stations. Unless an incoming SCA signal is trapped out, interference with the L and R signals is likely to occur. An SCA signal lacks the high fidelity of the main FM transmission, and is used chiefly in lieu of "piped music" in commercial establishments.

9.4 BASIC GENERATOR APPLICATION

One of the fundamental tests provided by a stereo-multiplex generator is a check of *separation* in operation of a stereo-multiplex unit or adapter, or of a complete receiver. That is, if we apply an L signal to an adapter, we might expect to obtain an output from the left channel only; or, if we apply an R signal to the adapter, we might expect to obtain an output from the right channel only. In theory, this is true; but, in practice, separation is less than perfect. If we apply an L signal with an amplitude that provides maximum rated output from the left channel, a receiver will often be considered to have acceptable separation if the L output from the right channel is 20 or 30 dB down.

Figure 9–17 shows a test arrangement for a separation measurement. The adapter or receiver is driven by a stereo-multiplex generator. Left and right channel outputs are measured with a VTVM or calibrated oscilloscope. If an adapter is driven directly, a composite audio signal is applied from the generator. On the other hand, if the adapter is driven through an FM tuner, the RF output from the generator is applied. If a VTVM with a dB scale is utilized, the amount of separation can be read directly by connecting the instrument in turn to each channel, and subtracting the two readings. The difference between the readings is equal to the amount of separation in dB units.

With reference to Fig. 9–10, separation entails a cancellation of +L and −L signals, and cancellation of +R and −R signals. Acceptable cancellation requires that the bandpass circuit have correct frequency limits, and that it does not introduce excessive phase shift into the L − R sideband signal. Although phase shift cannot be completely eliminated, it can be minimized by aligning for a flat pass characteristic and correct cut-off frequencies, as depicted in Fig. 9–11. Acceptable cancellation also requires

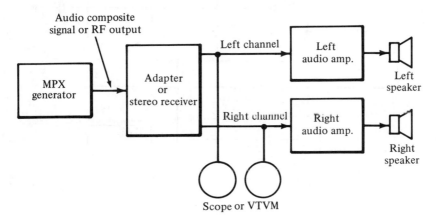

Fig. 9–17   Test arrangement for measurement of separation

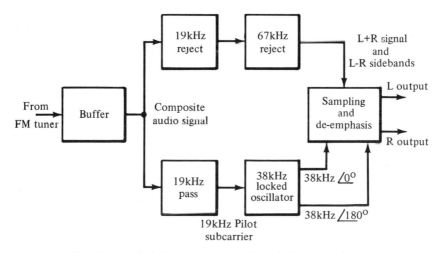

Fig. 9–18   Block diagram of a switching-bridge multiplex unit

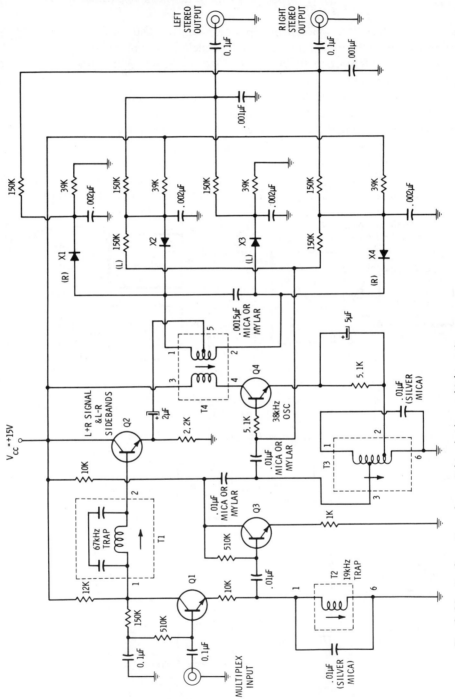

Fig. 9–19 Configuration of a switching-bridge multiplex unit

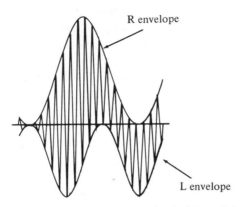

Fig. 9–20 Composite signal with individual L and R modulation envelopes

closely controlled signal levels, so that the subtractive process is as nearly exact as possible. Therefore, this type of multiplex adapter is provided with a separation control which is adjusted by the operator for optimum separation.

One of the most widely used multiplex configurations is the switching-bridge type, shown in block form in Fig. 9–18. The corresponding circuit diagram is shown in Fig. 9–19. It is basically a sampling arrangement which develops the upper and lower envelopes of the reconstituted stereo signal. That is, the composite stereo waveform comprises an R-signal envelope on one side of the axis, and an L-signal envelope on the other side of the axis, as exemplified in Fig. 9–20. In turn, 38-kHz samples are taken of the L and R envelopes, and the L and R signals are thereby separated. This is the basic function of the arrangement shown in Fig. 9–18.

Note in Fig. 9–19 that Q1 has a high input impedance, due to its substantial emitter degeneration. The 19-kHz pilot subcarrier signal is dropped across T2, which operates as a 19-kHz resonant trap. Thus, the pilot subcarrier signal is fed to Q2, but does not appear in the collector circuit due to the very high amount of degeneration at 19 kHz. Transistor Q2 operates as an emitter-follower, and serves as a buffer stage between the input stage and the subcarrier oscillator. Transistor Q3 also serves as a buffer, and amplifies the 19-kHz pilot-subcarrier signal. In turn, Q4 operates as a 38-kHz locked oscillator and is synchronized by the output from Q3. Semiconductor diodes CR1 through CR4 operate in a switching-bridge configuration. Diodes D1 and D4 provide R-signal output, whereas diodes D2 and D3 provide L-signal output. The de emphasis networks comprise series 150K resistors and 0.001-μF parallel capacitors.

A switching-bridge multiplex unit will provide a separation of approximately 25 dB between the L and R channels. Note that an input signal of

0.5 volts rms, or more, is necessary to keep the 38-kHz subcarrier oscillator tightly locked. If the oscillator jumps in and out of sync lock, serious noise and distortion occur. Although there is some residual feedthrough of 19-kHz and 38-kHz voltages in the configuration of Fig. 9–19, the filter action provides a rejection of more than 20 dB. Substantial feedthrough of residual voltages can cause trouble by overloading the following audio-amplifier stage.

As illustrated in Fig. 9–20, the reconstituted signal has different upper and lower modulation envelopes. The upper envelope is the R signal waveform, and the lower envelope is the L signal waveform. This method of developing the individual R and L signals is widely employed in envelope detector receiver design, utilizing a pair of oppositely polarized diodes.

QUESTIONS AND PROBLEMS

True-False

1. Stereophonic sound reproduction entails a pair of signals that basically involve a pair of microphones and a pair of speakers.
2. An FM multiplex generator is a simple form of transmitter.
3. Multiplexing is used to generate two signals from one signal.
4. An FM stereo-multiplex generator is a simplified form of transmitter.
5. The stereo-multiplex signal cannot be received with a standard FM receiver.
6. The 38-kHz subcarrier is transmitted with the FM signal.
7. The involved multiplexing system is used to provide high-fidelity stereophonic sound reproduction in an FM channel that normally accommodates only a high-fidelity monophonic signal.
8. The FM stereo-multiplex generator in Fig. 9–13 is a fixed-tone transmitter.
9. One of the fundamental tests provided by a stereo-multiplex generator is to check separation of the L + R and L − R signals.
10. Twenty to thirty dB down is acceptable separation between channels of an FM stereo-multiplex receiver.
11. When the adapter is driven through an FM receiver, the FM output from the generator is used.
12. The separation control is adjusted to give maximum channel separation.
13. The pilot subcarrier signal has a frequency of 38-MHz.

Multiple-Choice

1. A single audio signal occupies _____ of an FM channel.
   (a) ½ of the bandwidth
   (b) ¼ of the bandwidth
   (c) almost all the bandwidth

2. The maximum deviation of an FM carrier results in a frequency variation of
   (a) ±25 kHz.
   (b) 75 kHz.
   (c) ±75 kHz.
3. The purpose of the 38-kHz subcarrier is to
   (a) cancel the 19-kHz signal.
   (b) encode the L − R signal.
   (c) generate the high frequencies that cannot be broadcast in the allowed bandwidth.
4. The 38-kHz subcarrier is suppressed at the transmitter
   (a) to improve the signal to noise ratio.
   (b) because it carries no information.
   (c) to reduce the bandwidth.
5. The pilot subcarrier in an FM stereo-multiplex generator is made variable to
   (a) reduce distortion.
   (b) increase the separation.
   (c) test the receiver synchronization capability.
6. Acceptable separation of the L output from the right channel is _____ down.
   (a) 5 to 10 dB
   (b) 10 to 20 dB
   (c) 20 to 30 dB
7. The amount of separation between channels can be read by using a TVM on the _____ scale.
   (a) DC
   (b) AC
   (c) dB
8. If the multiplex adapter is driven directly, the signal used is
   (a) an FM signal.
   (b) a composite audio signal.
   (c) an RF output signal.
9. The pilot subcarrier signal is used to develop
   (a) separation bias.
   (b) signal sync.
   (c) the 38-kHz oscillator signal.

General

1. Describe a basic stereophonic sound system.
2. Explain how two signals can occupy the same channel without mutual interference.
3. What is the compatability requirement for an FM multiplex signal?
4. What is the purpose of the 38-kHz subcarrier?
5. What is the frequency range of the encoded L − R signal?
6. What is the purpose of the phase inverter following the AM detector in the circuit shown in Fig. 9–10?
7. Why is the 38-kHz subcarrier suppressed at the FM stereo-multiplex transmitter?

8. What is the purpose of the 19-kHz pilot subcarrier?
9. How is the 38-kHz subcarrier reconstituted at the receiver?
10. Basically, what is an FM stereo-multiplex generator?
11. What is the purpose of the 67-kHz test signal provided by the generator shown in Fig. 9–16?
12. Explain how a separation test is applied to a stereo-multiplex receiver.
13. Explain the separation test circuit depicted in Fig. 9–17.
14. How does the bandpass and matrix-type multiplex adapter differ from the switching-bridge type of multiplex adapter?

10

White-Dot
and Crosshatch
Generators

10.1 INTRODUCTION

White-dot and crosshatch generators are specialized types of rectangular wave generators. They are used to produce TV receiver screen patterns as shown in Fig 10–1. These patterns serve as a reference indication in making maintenance adjustments on color picture tubes. Instruments in this classification always provide white patterns on a dark background. In this respect they differ from linearity pattern generators which may provide black patterns on a white background. White-dot and crosshatch patterns are commonly called *convergence patterns* because they are utilized in picture-tube convergence procedures. Variations of the basic patterns illustrated in Fig. 10–1 include vertical lines, horizontal lines, single vertical and/or horizontal lines, a single dot, or combinations of crosshatch and dot patterns such as shown in Fig. 10–2.

Most convergence generators provide both crosshatch and dot signals, and are also combined with color-bar generators. Provision of these functions in a single instrument reduces the cost of equipping a service shop. Many generators supply a modulated RF output only, although a video-frequency output is also provided by some instruments. If a video-frequency output is available, it facilitates signal-injection tests during receiver troubleshooting procedures. Nearly all convergence generators are solid state and operate from self-contained batteries. Since portability is a basic requirement, compactness and rugged construction are essential design considerations.

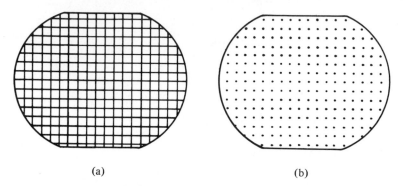

(a)   (b)

Fig. 10–1 Television test patterns: (a) crosshatch pattern; (b) dot pattern (Courtesy of Hickok Elec. Inst. Co.)

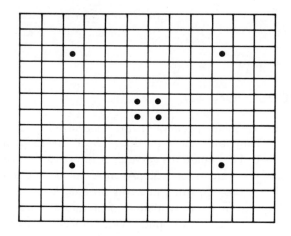

Fig. 10–2 Crosshatch pattern with dots (Courtesy of B & K Mfg. Co.)

10.2 BASIC PATTERN SIGNALS

Pulse waveforms are utilized to produce both dot and line patterns. For example, let us consider a signal waveform that will produce a vertical white line down the center of the picture-tube screen. With reference to Fig. 10–3, a pulse is inserted half-way between horizontal sync pulses, on each successive scan. The pulse extends from the black level to the white level, and produces a white dot by "turning on" the scanning beam as it reaches its half-way point between screen edges. It is evident that a vertical white line is displayed when white dots are lined up progressively down the screen. The width of the line depends on the width of the pulse. Note that vertical sync pulses are not required in this example, because the vertical line display is unchanged by "vertical rolling."

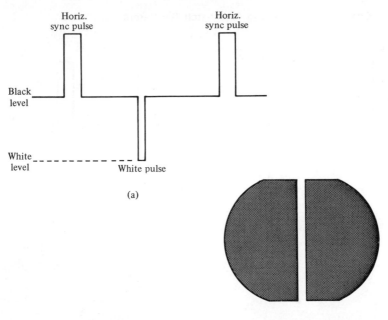

(a)

(b)

Fig. 10-3 Single pulse per trace displays a vertical white line: (a) video waveform repeated on each horizontal scan; (b) resulting vertical white line picture-tube display

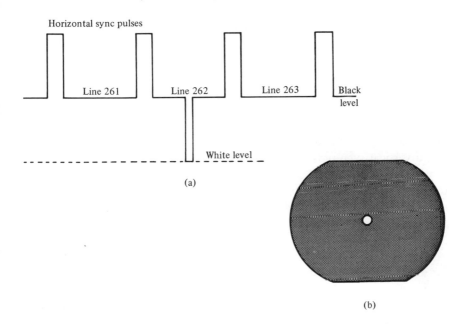

(a)

(b)

Fig. 10-4 Display of a single white dot: (a) video-signal waveform; (b) picture-tube display

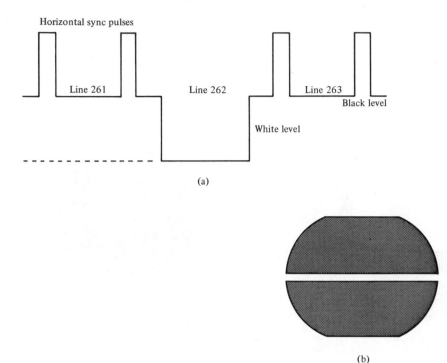

(a)

(b)

Fig. 10–5   Display of a single horizontal white line: (a) video-signal waveform; (b) picture-tube display

Next, to display a white dot at the center of the picture-tube screen, as shown in Fig. 10–4, the pulse is inserted in only one scanning line. That is, the signal comprises only horizontal sync pulses and black levels up to the 262nd line. On the 262nd line, a "white" pulse is inserted, which produces a white dot in the center of the screen. The remaining lines from 263 to 525 comprise only horizontal sync pulses and black levels. Note that a vertical sync pulse must also be included in the signal of Fig. 10–4, to maintain the dot at a position half-way between the top and bottom of the screen. This vertical sync pulse is inserted in the signal every 1/60 second. The vertical sync pulse is a broad rectangular pulse, and it occupies an interval equal to 30 horizontal scanning intervals. In turn, 495 of the 525 lines are actually visible on the picture-tube screen.

To display a single horizontal line on the picture-tube screen, the foregoing signal is modified by employing a 60-μsec pulse instead of a 0.25-μsec pulse, as shown in Fig. 10–5. In other words, a horizontal line occupies an elapsed time of 63.5 μsec, of which the active time is somewhat less due to the width of the horizontal sync pulses. A time of 5.1 μsec

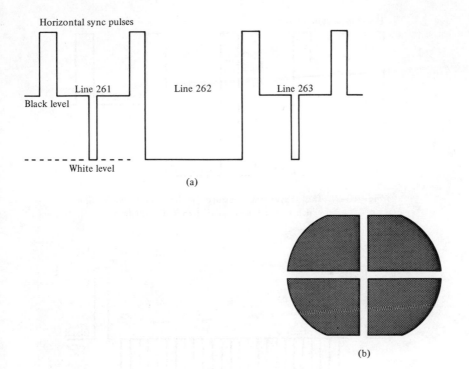

Horizontal sync pulses

Line 261   Line 262   Line 263

Black level

White level

(a)

(b)

Fig. 10–6   Display of a vertical and a horizontal white line:
(a) video-signal waveform; (b) picture-tube display

is lost on each horizontal-retrace interval. As in the foregoing example, the horizontal line will not be locked at screen center unless we also insert a vertical sync pulse into the signal each 1/60 second. Note that the horizontal scanning frequency is 15,750 Hz, and the vertical scanning frequency is 60 Hz.

When we display a vertical and a horizontal line through the center of the picture-tube screen, as shown in Fig. 10–6, we employ a combination of the waveforms shown in Fig. 10–3 and Fig. 10–5. As seen in Fig. 10–6, a narrow pulse is inserted in each horizontal interval, with the exception of the 262nd line which has a 60-μsec pulse. A vertical sync pulse is required in the signal to maintain the position of the horizontal white line on the picture-tube screen. Designers of white-dot and crosshatch generators often utilize dots and lines which include two scanning lines instead of one. Thereby, the pattern is made more visible although its structure is not as fine. For example, the waveform in Fig. 10–7 provides a horizontal white line with a thickness of two scanning lines.

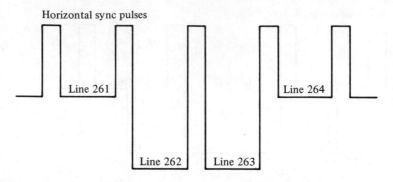

Fig. 10–7   This waveform produces a horizontal white line which has a thickness of two scanning lines

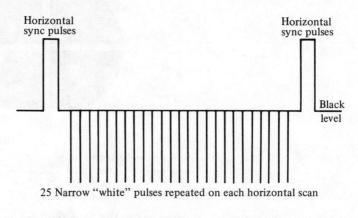

Fig. 10–8   Waveform for producing 25 vertical white lines

In case it is desired to display 15 horizontal lines on the picture-tube screen, successive lines will be separated by 33 horizontal scanning intervals, assuming that the raster comprises 500 scanning intervals. That is, the wide pulse depicted in Fig. 10–5 would be flanked by 33 scans at black level, followed by one scan with a wide pulse. This sequence is repeated 15 times to display a horizontal line pattern of 15 lines. Next, to add vertical white lines into the pattern, we employ a narrow pulse waveform as shown in Fig. 10–3. If we wish to produce 25 vertical lines, the signal will comprise 25 pulses between consecutive sync pulses, as shown in Fig. 10–8. The complete signal, comprising the foregoing wide and narrow pulses, will then display a white-line pattern consisting of 15 horizontal lines and 25 vertical lines.

Note that the waveform shown in Fig. 10–7 is developed when a simplified scanning system, which does not provide interlacing, is used. Many pattern generators employ this arrangement; it is sometimes referred to as "random interlace." The more elaborate generators employ locked horizontal and vertical oscillators that provide accurately interlaced scanning fields. In such case, the even lines are scanned first, and then the odd lines are scanned. Accordingly, the wide pulses modulating the signal on lines 262 and 263 in Fig. 10–7 must be separated by a time interval of 1/60 second when interlaced scanning is utilized.

10.3 SOLID-STATE WHITE-DOT AND
CROSSHATCH GENERATOR

The solid-state white-dot and crosshatch generator illustrated in Fig. 10–9 provides crystal-controlled test signals for dot, crosshatch, horizontal line, vertical line, gray-scale, and color-bar patterns. The color-bar section will be more fully considered in the next chapter. With reference to the block diagram shown in Fig. 10–10, combinations of logic circuits are employed to generate various pattern signals. A stable sine wave is generated by a master clock oscillator. This signal is shaped by succeeding stages, and proceeds to a divider chain that consists of a series of flip-flop multivibrator circuits. These flip-flops divide the frequency of the master clock signal.

At various points in the divider chain, signals of required frequency are picked off and applied to other logic circuits consisting of AND, NAND, NOR, or OR gates. These circuits, in turn, combine the proper signals to produce the desired output pattern signals. Note that the master clock oscillator generates the primary signal from which all other signals in the

Fig. 10–9 Solid-state white dot and crosshatch generator
(Courtesy of Heath Co.)

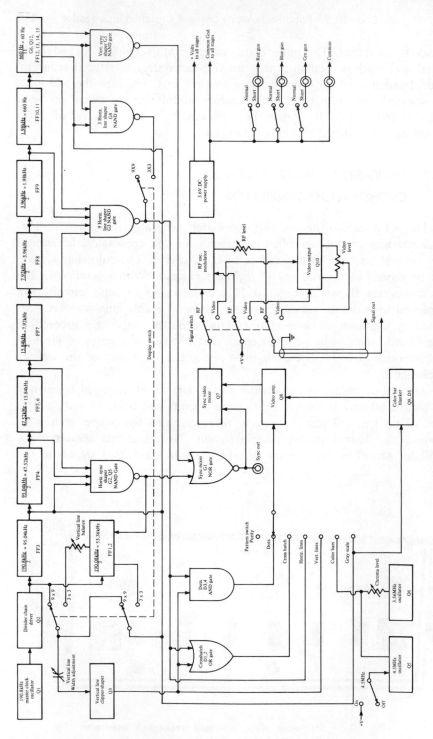

Fig. 10–10 Block diagram of generator (Courtesy of Heath Co.).

timing circuits are obtained (Fig. 10–10). To provide accurate timing signals, the output of the master clock must be very stable. This is accomplished by using a crystal-controlled Pierce oscillator which generates a 190.08-kHz signal.

10.4   BASIC LOGIC GATES

It follows from Fig. 10–10 and the foregoing discussion that operation of the generator involves more than flip-flop circuits. That is, the various gates (logic gates) serve to determine when to perform an operation, what operation to perform, and in which of several ways an operation will be performed. We call this type of circuit action *machine* or *computer* logic. The logic patterns with which we are concerned in this situation are termed AND, OR, NOR, and NAND functions. These functions are performed by means of semiconductor switching circuits. Although machine logic can be compared with arithmetical operations, there is considerable distinction between them. In elementary arithmetic, the four basic operations are addition, subtraction, multiplication, and division. In machine logic, there are three basic operations: AND, OR, and NOT.

For convenience, familiar arithmetic symbols are used for two of the three machine-logic operations. Note that:

AB means A AND B;
A + B means A OR B.

In other words, the OR operation is indicated by the addition symbol; the AND operation is indicated by the multiplication symbol. If we choose, we can use any multiplication symbol to indicate the AND operation. For example,

(A + B)(C) means A OR B, AND C;
A · B means A AND B.

A bar is placed over a letter to indicate a NOT operation. Thus,

$\overline{A}$ means NOT A.
$A\overline{B}$ means A AND NOT B.

The term NOR is a contraction of NOT OR, and it signifies that the logical function performed is that of the OR gate, followed by an inverter or NOT circuit. An apostrophe symbolizes this operation. Thus,

(A + B + C)′ means NOR, the equivalent of NOT OR.

The term NAND is a contraction of NOT AND; it consists of an AND gate and an inverter or NOT gate. An apostrophe is also used to symbolize this operation. Thus,

$$(A \cdot B \cdot C)' \text{ means NAND, the equivalent of NOT AND.}$$

These terms will become clearer as we discuss the physical models for the various operations. Figure 10–11 shows a physical model for the AND operation. A simple electric circuit that lights a lamp is used for illustration. The AND operation is represented by series switching. In the open position, the switch condition may be denoted by 0. Conversely, in the closed position, the switch condition may be denoted by 1. We may also denote absence of current flow through the lamp by 0, and denote current

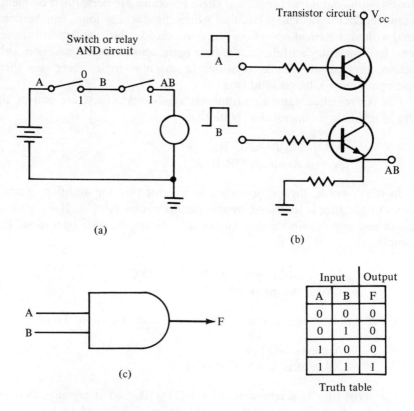

(a)

(b)

(c)

Input		Output
A	B	F
0	0	0
0	1	0
1	0	0
1	1	1

Truth table

(d)

Fig. 10–11 Basic AND circuit: (a) switch symbolism; (b) transistor circuit; (c) circuit diagram with symbol used to denote an AND circuit; (d) truth table

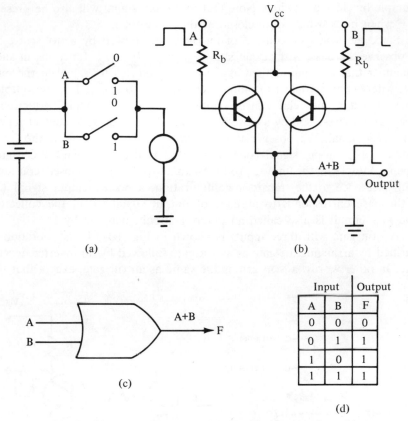

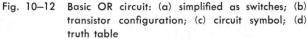

(c)

Input		Output
A	B	F
0	0	0
0	1	1
1	0	1
1	1	1

(d)

Fig. 10–12   Basic OR circuit: (a) simplified as switches; (b)
transistor configuration; (c) circuit symbol; (d)
truth table

flow by 1. In turn, a *truth table* may be compiled as shown, to tabulate
the combinations of various conditions and results.

We recognize that the AND circuit operates in terms of multiplication.
If a switch is open, it will be tabulated in the truth table as 0. Conversely,
if a switch is closed, it is tabulated in the truth table as 1. Next, an 0
in the output column indicates that no current flows; on the other hand, a
1 in the output column indicates that current will flow. In summary, this
truth table shows us that there is only one condition of the switches that
will permit current flow through the lamp. This condition occurs when
switches A and B are both closed at the same time.

Proceeding to the OR gate configuration, we observe in Fig. 10–12 that
an OR circuit can be represented electrically by means of two single-pole
switches connected in parallel. There is an output signal present at F when
an input signal is applied at A or B. That is, a signal is applied in this

example by closing a switch. Note that an output signal will also be present at F when both switches are closed at the same time.

Since the NOR gate consists of an OR gate followed by a NOT gate, let us observe the basic NOT circuit shown in Fig. 10–13. It functions in such a manner that a signal applied at the input produces no signal at the output, whereas no signal at the input corresponds to a signal at the output. Thus, a switch and relay can be connected to form a simple NOT circuit. A signal is represented by a closed contact, whereas an open contact represents NO signal. We observe that a signal (closed contact) at the input produces NO output signal (open contact). On the other hand, NO signal (open contact) at the input, produces an output signal (closed contact). With reference to the transistor configuration, a positive input signal (1) at the base corresponds to absence of output signal (0) at the collector. The NOT circuit is also called an inverter circuit, denoted by I.

A NOR gate with three inputs is shown in Fig. 10–14. Its operation is clarified by arranging the gate as an OR gate followed by an inverter or NOT gate. In other words, a NOR gate is the same as an OR gate, except that the

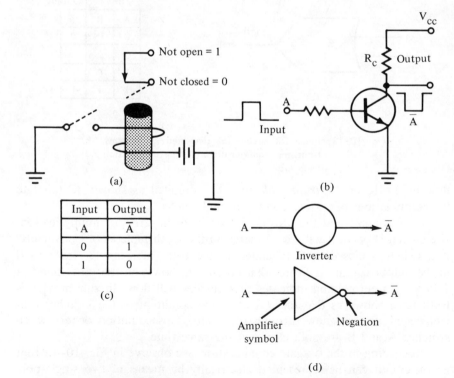

Fig. 10–13 Basic NOT circuit: (a) as a relay; (b) transistor amplifier; (c) truth table; (d) circuit symbol

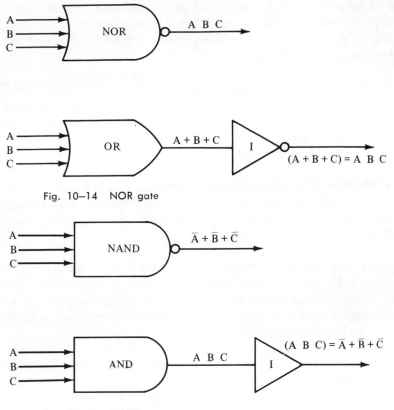

Fig. 10–14   NOR gate

Fig. 10–15   NAND gate

output is inverted; an output signal is present only when A, B, and C are simultaneously 0. A NAND or NOT AND gate is shown in Fig. 10–15.

10.5   GENERATOR ORGANIZATION

The divider chain driver circuit shapes the 190.08-kHz sine wave into a square-wave. The repetition rate of the square-wave is divided down for driving the gate circuits and the output switching circuits. Signals from the various gate circuits, driver, shaper, and crystal-controlled oscillator circuits are coupled to the pattern selector switch. Each position of this switch selects a signal of proper frequency and waveshape to produce a particular pattern. A brief description of these signals follows:

Vertical lines. The output from the master clock and divider chain driver passes through the clipper-shaper stage which differentiates the waveform. A pulse is thereby formed that will develop a narrow vertical line on the picture-tube screen. This pulse is combined with the sync signal in the sync/video circuit. In turn, this combination forms the required number of pulses per horizontal sweep period. Thereby, the desired number of vertical lines are displayed.

Horizontal lines. The output signal from the divider chain is fed to a NAND gate. This results in a pulse that displays a narrow horizontal line on the picture-tube screen. The pulse is combined with the sync signal in the sync-video circuit. This combination forms the required number of pulses per vertical scan period.

Crosshatch. The horizontal and vertical line signals are combined in an OR gate. Both the vertical and horizontal signals from the output of the OR gate produce the vertical and horizontal lines which form the crosshatch pattern on the picture-tube screen.

Dots. An AND gate combines the vertical and horizontal signals to produce a dot pattern. Since both signals must be present to produce a pulse output, this condition exists only when the vertical and horizontal lines cross. Thereby, horizontal and vertical rows of dots are formed. Details are explained subsequently.

Gray scale. The gray-scale pattern is produced by combining pulses to form the vertical and horizontal bars in a wide-bar crosshatch pattern. The half-frequency signal in each direction superimposes every other bar to display a change in gray shade.

10.6   FUNCTIONAL DESCRIPTION

In Fig. 10–10, the 63.36-kHz signal from FF1, FF2, and the 240-Hz signal from FF12 are combined to produce the display that is called the 3×3 display. Next, the 190.08-kHz signal from Q2 and the 660-kHz signal from FF11 are combined to produce the display called the 9×9 display. Note that the combined vertical and horizontal sync signals and the pattern signal from the video amplifier are combined in the sync/video stages to provide a composite video signal. This composite signal is then coupled to either the video-output stage or to the RF oscillator stage, depending on the setting of the signal switch.

The video-output stage in Fig. 10–10 supplies the composite video signal for coupling directly into the video circuits of a TV receiver. Note that the RF oscillator circuit produces an RF carrier that can be modulated by the video signal, and which is tunable through TV channels 2 to 6, inclusive. A flip-flop, or bistable multivibrator, can be compared with a two-position switch. Flip-flops are manufactured in many configurations and with several special features. A popular version that has great versatility is termed the J-K flip-flop; this is the type employed in the generator under discussion. Any FF has two outputs; these are denoted by Q and $\overline{Q}$. The symbol $\overline{Q}$ has the meaning of "not Q." The state of an FF corresponds to the level of Q.

A J-K FF has three input terminals. These are called set (S), clear (C), and trigger (T). They can be utilized in such manner that a pulse applied to the trigger input will or will not switch the output, depending on the ON conditions at the set and clear terminals. In addition, a preset (P) terminal provides a way to return the FF to a particular state, independently of trigger input. With both set and clear grounded, the output becomes ON once each time that two pulses have been applied to the trigger input. This is termed a divide-by-two action, and it is independent of frequency as long as it does not exceed the switching speed limit of the FF.

Figure 10–16 shows how flip-flops FF1-FF2 are connected to divide-by-three, and the input and output waveform relationships. Assume that the Q output of flip-flop FF1 is at a 1 level and the Q output of flip-flop

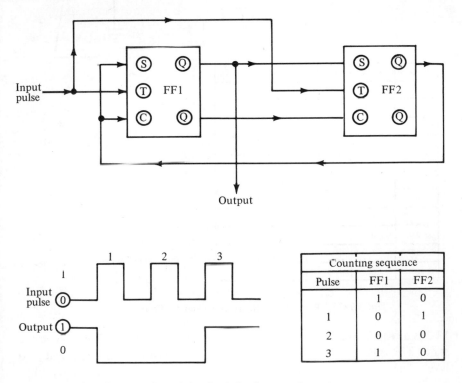

Fig. 10–16 Plan of the divide-by-three configuration

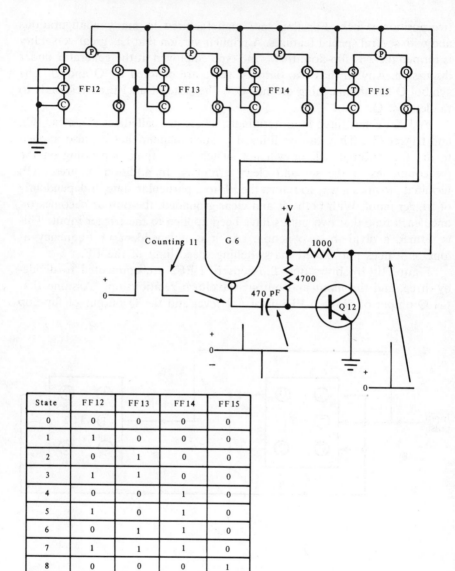

Fig. 10–17 Plan of the divide-by-eleven configuration

State	FF 12	FF 13	FF 14	FF 15
0	0	0	0	0
1	1	0	0	0
2	0	1	0	0
3	1	1	0	0
4	0	0	1	0
5	1	0	1	0
6	0	1	1	0
7	1	1	1	0
8	0	0	0	1
9	1	0	0	1
10	0	1	0	1

FF2 is at the 0 level. The first trigger pulse causes both flip-flops to change state. This places the set (S) and clear (C) inputs of flip-flop FF2 at a 0 level. The second trigger pulse causes flip-flop FF2 to change state. Flip-flop FF1 was prevented from switching by the 1 level at its set and clear inputs.

Note that the trigger pulse appears at the trigger (T) inputs of both flip-flops simultaneously. By the time that flip-flop FF2 changes state, the pulse no longer affects flip-flop FF1. The third trigger pulse will cause flip-flop FF1 to change state, but flip-flop FF2 will remain the same. The reason for this is due to 0 level appearing at the Q output of FF2, which matches the 0 level (prior to flip-flop FF1 switching) at the set input. This relationship is shown in the counting sequence table in Fig. 10–16.

Next, Fig. 10–17 shows how flip-flops FF12-FF13-FF14-FF15 are connected to divide-by-eleven, with the input and output waveform relationships. When a 1-level signal is applied to the preset inputs, the Q output of each flip-flop is forced to the 0 level. Since the two outputs are always at opposite levels, the $\overline{Q}$ output will be at the 1 level. Gate G6 will produce a positive square-wave pulse only when all four input terminals to G6 are at the 0 level, which occurs only on the eleventh pulse (or the tenth state). The square-wave pulse at the output of G6 is differentiated by the 470-pF capacitor and the 4700-ohm resistor, which produces a positive pulse on the leading edge and a negative pulse on the trailing edge of the square wave. This negative pulse serves to cut off Q12 and causes a rise in collector voltage. This rise in voltage is applied to the preset inputs and sets the flip-flops in their preset state again, thus dividing down by eleven.

A gate is so named because it allows signals to pass only under controlled conditions. Four types of gates are employed in the system of Fig. 10–10. They are AND, NAND, OR, and NOR gates. The AND gate requires that all input signals be present before it will produce an output signal. The output signal from an AND gate has the same phase as its input signal, as may be seen in Fig. 10–18. Next, we observe that a NAND gate is a variation of the AND gate, in that the output signal is inverted when a NAND gate is

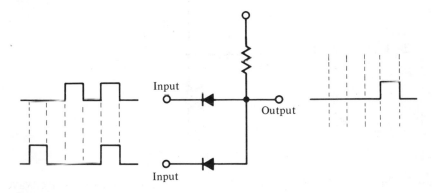

Fig. 10–18   Basic arrangement of an AND gate

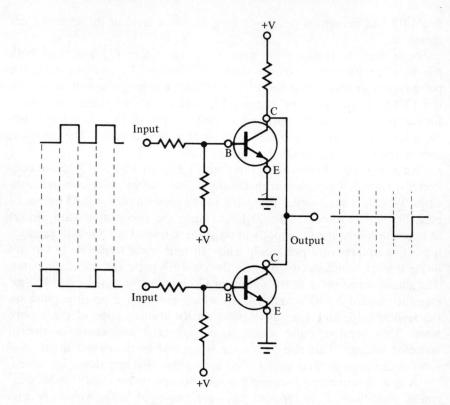

Fig. 10–19   Basic arrangement of a NAND gate

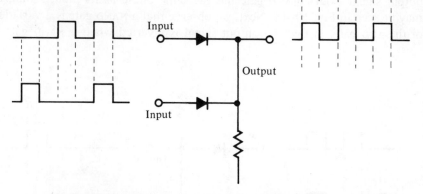

Fig. 10–20   Basic arrangement of an OR Gate

used. This fact is evident in Fig. 10–19 because the collector voltage wave-form is inverted with respect to the base voltage waveform.

An OR gate produces an output when one or the other of its input signals is present. The output from the OR gate is the same in phase as its input signal, as we observe in Fig. 10–20. Next, a NOR gate is an OR gate that inverts the output signal, as we observe in Fig. 10–21. Diodes D1 and D2 in Fig. 10–10 function as an OR gate which combines the horizontal and vertical line pulses into the crosshatch pattern. Diodes D3 and D4 function as an AND gate. An output results only when both horizontal and vertical line pulses are present, which occurs at the crossover points. Thus, the dot pattern is formed.

The "Quad-two-input gate" IC comprises four pairs of transistors. Each pair forms a two-input gate. The output is inverted and the gate may be used for the NAND or NOR function. One such gate is G1, which functions as a NOR gate. In this gate, the horizontal and vertical sync pulses are combined to form the composite sync signal. Another gate, G5, functions as a NAND gate. Two input pulses of different repetition rates and pulse duration are combined to give an output only during the brief moment that both inputs are present. Gate G3 uses two pairs of transistors tied in parallel to

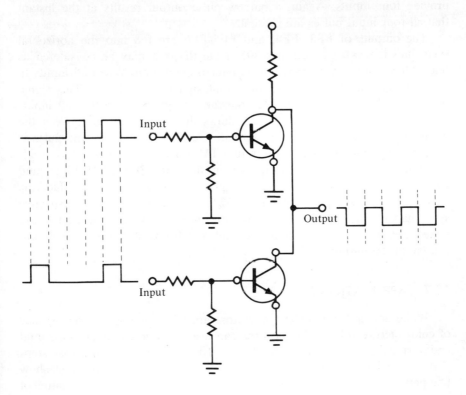

Fig. 10–21 Basic arrangement of a NOR gate

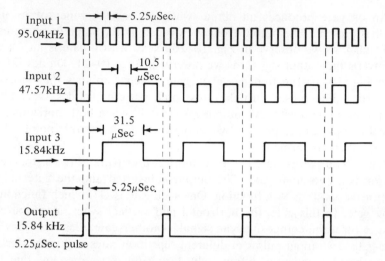

Fig. 10–22   Composite sync signal formation

provide four inputs. Again, a narrow pulse output results at the instant that all four input pulses are present.

The outputs of FF3, FF4, and FF5-FF6 are fed into the horizontal sync shaper, NAND gate G2 and Q5. Note that Q5 may be considered as one added input to G2, since the outputs are common. When all inputs to G2 and Q5 are at zero level, an output signal is produced. This signal remains at the output terminal for the same length of time that all inputs are at the zero level. Figure 10–22 shows the time relations between the input signals at the output of NAND gate G2. The 5.25-μsec sync pulse is similar to the horizontal sync pulse in a TV transmitter signal.

In the same manner, the outputs from flip-flops FF9, FF10-FF11, and FF15 are fed into the vertical sync shaper (NAND gate G5) to produce the 60-Hz vertical sync pulse which is 252 μsec wide. From their respective sync shapers, the signals are fed to NOR gate G1, where the signals are mixed and form the composite (vertical and horizontal) sync signals appearing at the output of G1.

10.7   APPLICATIONS

White-dot and crosshatch generators are used chiefly for convergence of color picture tubes. The patterns can also be used to check horizontal and vertical deflection linearity. Figure 10–23 shows the principal steps that are involved in convergence procedure. Although the illustrations show line patterns, dot patterns can also be employed. It is basically a matter of personal preference whether dots or lines are used. Convergence controls

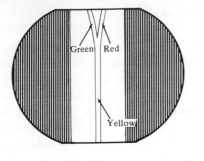

(a)

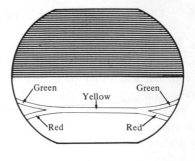

(b)

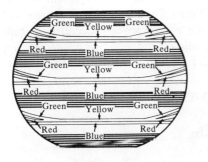

(c)

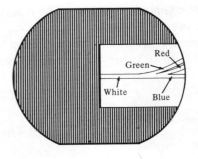

(d)

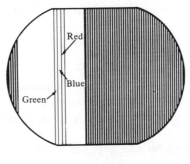

(e)

Fig. 10 23 Progressive convergence of a color picture tube: (a) vertical-dynamic controls are adjusted to converge the red and green lines at top and bottom; (b) horizontal-dynamic controls are adjusted to converge the red and green bars at the left and right edges of the screen; (c) horizontal-bar pattern; (d) right-edge horizontal-bar pattern; (e) vertical-bar pattern

are grouped into static and dynamic functions. The static convergence controls are adjusted to obtain correct convergence at the center of the screen. On the other hand, the dynamic controls are adjusted to obtain correct convergence at the top, bottom, left, and right edges of the screen.

Misconvergence is indicated by the splitting up of white lines or dots into colored lines or dots. That is, when red, green, and blue lines are superimposed, white lines are formed. At the start of the convergence procedure, we usually obtain white lines or dots only near the center of the screen. However, as the dynamic adjustments are progressively made, the white lines or dots are extended farther out toward the screen edges. Since dynamic and static controls tend to interact, readjustment of the static controls is occasionally required. Finally, the entire screen area will be occupied by white lines or dots, and the convergence procedure is completed. Readers who are interested in details of convergence procedure are referred to color-TV receiver service data.

QUESTIONS AND PROBLEMS

True-False

1. White-dot and crosshatch generators always produce white patterns on dark backgrounds.
2. White-dot patterns and crosshatch patterns are called convergence patterns.
3. Sine waveforms are utilized to produce both dot and line patterns.
4. Vertical sync pulses are not required when a vertical line is made up of vertical dots.
5. White-dot and crosshatch generators are used to test the linearity of both the vertical and horizontal patterns.
6. The logic terms A + B mean A OR B.
7. A truth table gives all results of all circuit modes.
8. An absence of an output signal is denoted by a 1.
9. In Fig. 10–10, the crosshatch pattern is taken from the output of the OR gate.
10. The output signal from an AND gate has a 180° phase relationship to the input signal.
11. The static convergence controls are adjusted to obtain correct convergence at the edges of the screen.
12. Dynamic and static controls tend to interact.

Multiple-Choice

1. White-dot and crosshatch patterns are commonly called _____ patterns.
   (a) convergence
   (b) linearity
   (c) background

2. Both dot and line patterns are produced by
   (a) sine waves.
   (b) pulse waveforms.
   (c) sawtooth waves.
3. The horizontal scanning frequency is _____ Hz and the vertical scanning frequency is _____ Hz.
   (a) 30; 60
   (b) 60; 15,750
   (c) 15,750; 60
4. In Fig. 10–10, the timing signals are developed by
   (a) the 60-Hz line frequency.
   (b) a crystal-controlled oscillator.
   (c) a free-running 60-Hz oscillator.
5. The term NOR means:
   (a) neither A nor B
   (b) not OR
   (c) the same as NAND
6. A table of all circuit conditions and results is called a
   (a) tabulator.
   (b) code table.
   (c) truth table.
7. White-dot and crosshatch generators are used chiefly for
   (a) convergence of color picture tubes.
   (b) linearity of color picture tubes.
   (c) intensity tests of color picture tubes.
8. Dynamic convergence controls are used to adjust the convergence at the _____ of the screen.
   (a) center
   (b) edges
   (c) center and edges

General

1. How do white-dot and crosshatch generators differ from linearity pattern generators?
2. Why are white-dot and crosshatch patterns commonly called convergence patterns?
3. Why do most convergence generators use solid-state circuitry?
4. Explain how a vertical white line, such as that shown in Fig. 10–3, can be produced on the screen of a television picture tube.
5. Explain how a dot can be displayed on the screen of a TV-picture tube.
6. Explain how a horizontal line, such as shown in Fig. 10–5, can be displayed on the screen of a TV-picture tube.
7. Explain how vertical and horizontal lines, such as illustrated in Fig. 10–6, can be displayed on the screen of a TV-picture tube.
8. Explain the term "random interlace" as it applies to the waveform in Fig. 10–7.
9. What is the function of the master clock oscillator in the block diagram shown in Fig. 10–10?

10. What is the frequency of the master oscillator circuit in the white-dot and crosshatch generator shown in Fig. 10–9?

11. Explain how each of the signals for the following patterns is produced in the generator shown in Fig. 10–10; vertical lines, horizontal lines, crosshatch, dots, and gray scale.

12. What is the purpose of the 470-pF capacitor and the 4700 kilohm (kΩ) resistor in the base circuit of Q12 in Fig. 10–17?

13. What is the basic definition of a gate circuit?

14. What is the difference between the AND circuit and the NAND circuit?

15. What is the difference between an OR circuit and a NOR circuit?

11

Color-Bar
Generators

11.1   INTRODUCTION

Two basic types of color-bar generators are in general use.* The keyed-rainbow type is preferred in the service trade, whereas the NTSC type of generator is utilized by most laboratories and by engineers in the color-television industry. A keyed-rainbow generator provides a linear phase sweep of the chroma subcarrier, as explained subsequently. This signal displays a color spectrum on the screen of a color picture tube; the spectrum of hues is called a rainbow pattern. To form a keyed-rainbow pattern, the linear phase sweep signal is modulated by a square wave in the generator, so that the rainbow pattern is divided into bars of color by successive black bars, as shown in Fig. 11–1.

A keyed-rainbow pattern provides chroma phase indications that facilitate analysis of circuit malfunctions and permit accurate adjustment of chroma-phase controls. The phases indicated in Fig. 11–1(b) will be explained in greater detail subsequently. Most keyed-rainbow generators supply a pattern of 10 chroma bars. However, a few generators supply a pattern of 3 bars, corresponding to the red, green, and blue chroma phases. An occasional generator is designed to provide single-bar displays corresponding to the red, blue, green, yellow, magenta, cyan, R-Y, and B-Y chroma phases. However, this type of generator is not as widely used as the other types.

Another basic type of color-bar generator provides an NTSC signal, as noted previously. NTSC is an abbreviation for National Television Standards Committee; this committee establishes the signal specifications for

---

* Color-television terminology and basic operational principles are explained in various reference texts such as *Color TV Training Course*, published by Howard W. Sams & Co., Indianapolis, Indiana.

color-television broadcast stations. An NTSC generator provides true primary and complementary color signals at full brightness and saturation. Figure 11–2 shows the screen pattern produced by a typical NTSC color-bar generator. The order of colors is arbitrary and often differs from one generator to another. In addition, most NTSC generators also provide

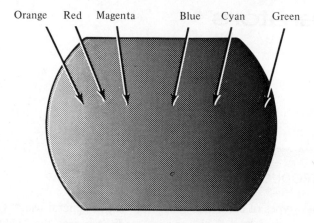

(a)

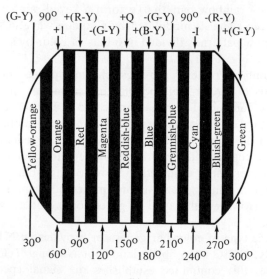

(b)

Fig. 11–1   (a) Unkeyed-rainbow pattern; (b) standard keyed-rainbow pattern

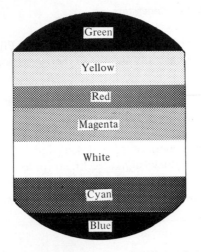

Fig. 11-2 Screen pattern produced by a typical NTSC generator

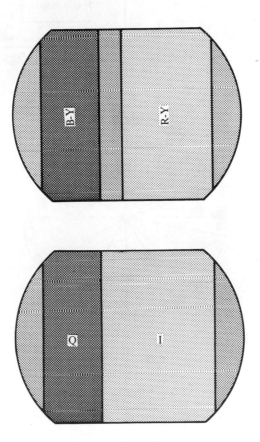

Fig. 11-3 Color-difference bar patterns

various color-difference patterns, as exemplified in Fig 11–3. These color-difference displays are basically similar to the corresponding keyed-rainbow bars in Fig. 11–1(b).

11.2 THE KEYED-RAINBOW SIGNAL

A keyed-rainbow generator provides a signal that is called an offset color subcarrier, a sidelock signal, a linear phase sweep, or a rainbow signal. The reason for these various terms will become clear after the signal characteristics have been explained. It is helpful to start with a discussion

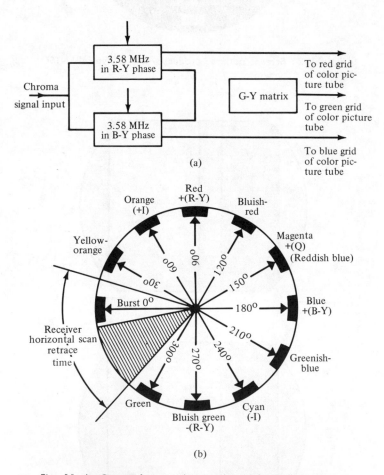

Fig. 11–4 Output from a chroma demodulator depends on the phase of the input chroma signal: (a) chroma demodulators are phase and amplitude detectors; (b) displayed hue depends on the phase of the chroma signal

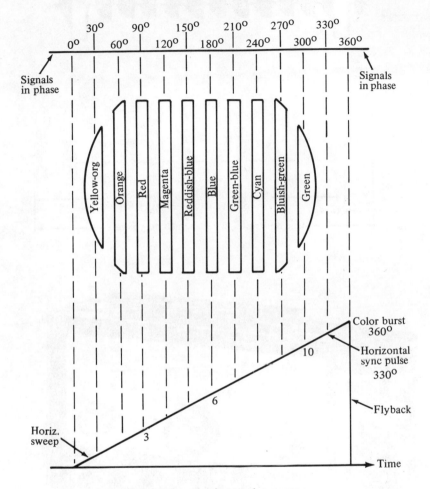

Fig. 11–5 Visualization of a linear phase sweep

of the unkeyed rainbow signal, sometimes called a color-display signal. It should be recognized that the chroma demodulators in a color-TV receiver are product detectors that demodulate an applied signal in accordance with its phase; the reference phase is established by the locally generated subcarrier voltage that is injected into the project detectors.

In other words, with reference to Fig. 11–4(a), full output from the R-Y demodulator is obtained when the input signal has a phase of 90° with respect to burst. Zero output is obtained from the R-Y demodulator when the input signal has a phase of 0° with respect to burst. Full output is obtained from the B-Y demodulator when the input signal has a phase of 0° with respect to burst. Zero output is obtained from the B-Y demodulator when the input signal has a phase of 90° with respect to burst. If the

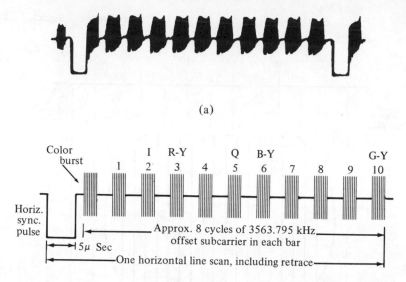

(a)

(b)

Color burst

Horiz. sync. pulse

I   R-Y   Q   B-Y   G-Y

1   2   3   4   5   6   7   8   9   10

5μ Sec

Approx. 8 cycles of 3563.795 kHz offset subcarrier in each bar

One horizontal line scan, including retrace

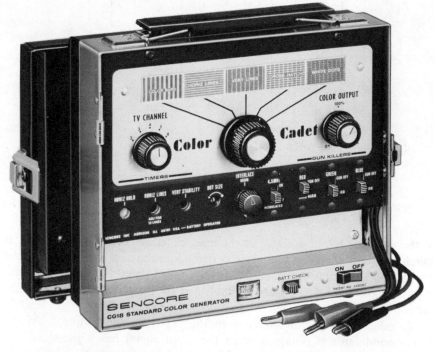

(c)

Fig. 11–6 (a) Keyed-rainbow signal waveforms; (b) signal characteristics; (c) appearance of typical keyed-rainbow generator (Courtesy of Sencore)

input signal has a phase of 120° with respect to burst, the output from the R-Y demodulator is 87% of maximum, and the output from the B-Y demodulator is 50% of maximum.

It is evident that the phase of the input chroma signal is swept (progressively changed) from 0° to 360° in Fig. 11–4, and the succession of colors indicated in (b) will be displayed on the picture-tube screen. That is, we will observe the chroma-bar pattern depicted in Fig. 11–1(b). Note that the frequency of the color burst is 3.579545 MHz (commonly rounded off and abbreviated as 3.58 MHz). Although a basic rainbow generator could process this frequency (the color-subcarrier frequency) through a phase modulator, design simplification is afforded by the circumstance that the color subcarrier is locally generated in a color-TV receiver, and injected into the chroma demodulators. Let us see how this fact bears on the design of a rainbow generator.

We will find that development of a rainbow pattern on the screen of a color picture tube can be accomplished by means of a CW signal which produces the same end result as phase modulation of the color subcarrier. This CW signal has a frequency of 3.563795 MHz, commonly rounded off and abbreviated as 3.56 MHz. This is the offset color-subcarrier frequency that was previously mentioned. An offset color subcarrier differs from the color-subcarrier frequency by 15,750 Hz, or it differs by the horizontal scanning frequency employed in the receiver. Now let us consider the circuit action that occurs in the R-Y demodulator when an offset color-subcarrier signal is applied.

It is evident that since the 3.56-MHz and the 3.58-MHz signals are both present in the R-Y demodulator, these two frequencies must beat together, and their difference frequency of 15,750 Hz will appear at the output of the R-Y demodulator. This beat output starts at zero, rises to a maximum, and decreases to zero at a 15,750-Hz rate. That is, the output signal is the same as if a phase-modulated 3.58-MHz signal had been applied to the demodulator. It is called a linear phase sweep, because the output beat frequency starts with the signals in phase, and progressively develops an increasing phase difference until 360° is reached, when the signals are once again in phase. Figure 11–5 is a visualization of a linear phase sweep. The waveform of a keyed-rainbow signal is shown in Fig. 11–6. Note that, due to the comparatively long flyback interval in a color-TV receiver, part or all of the first bar, the tenth bar, or both, may be invisible on the picture-tube screen.

11.3 WAVEFORM CHARACTERISTICS

We observe that 10 chroma bars are normally displayed in the pattern depicted in Fig. 11–5. On the other hand, the keyed-rainbow signal shown

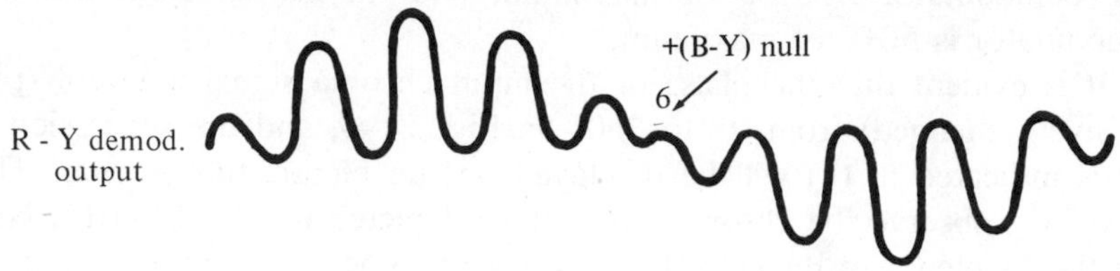

Fig. 11–7 (a) Output waveform from the R-Y demodulator;
(b) output waveform from the B-Y demodulator

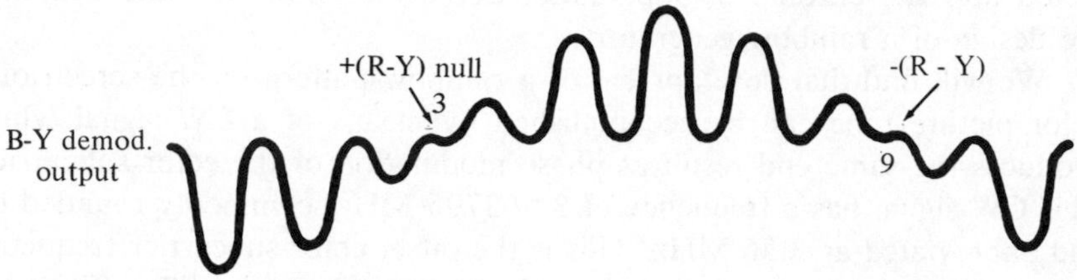

in Fig. 11–6 comprises 11 bursts. Note that the burst immediately following the horizontal sync pulse occurs at a time such that it is gated out and passes into the color-sync section of the receiver. That is, only 10 bursts proceed into the chroma section of the receiver, which includes the R-Y and B-Y demodulators shown in Fig. 11–4(a). In turn, the output wave-form from the R-Y demodulator comprises trains of 10 pulses with a sine-wave envelope, as shown in Fig. 11–7(a).

Note that the B-Y demodulator develops a similar output waveform, except that the phase is shifted 90° from that of R-Y demodulator output. Although it would be possible to design a keyed-rainbow generator to display more than 10 bars, it would not be useful for checking chroma circuits in color-TV receivers. The bandwidth of a chroma demodulator is approximately 0.5 MHz. In turn, 10 bars per horizontal scan (equivalent to 12 bars with blanking taken into account) operate the chroma demodulators at their maximum practical repetition rate. If we attempt to process 15 or 20 bars per horizontal scan, we find that the output waveforms from the chroma demodulators become objectionably distorted. Therefore, keyed-rainbow generators have been standardized with the signal characteristics shown in Fig. 11–6.

Note that the fundamental frequency of the waveform in Fig. 11–6 is equal to the horizontal scanning frequency multiplied by the total number of equivalent bars. In other words, the fundamental frequency of the waveform is equal to 15,750 × 12, or 189 kHz. It follows that a keyed-rainbow generator must employ an oscillator to generate this frequency; whether the 189-kHz oscillator is suitable for operation as a master oscillator necessarily depends on whether the vertical sync-pulse repetition rate is harmonically related in 189 kHz. Since 60 divides evenly into 189,000, it is evident that a 189-kHz oscillator may be employed as a master oscillator in a keyed-rainbow generator.

11.4   KEYED-RAINBOW GENERATOR CIRCUITRY

With reference to Fig. 11–8, the master oscillator in the generator is a crystal-controlled 189-kHz configuration. Crystal control is necessary in order to obtain the required frequency stability. The output waveform from the master oscillator is passed through a buffer stage which also serves as a waveshaper to produce a pulse-type output. Each 12th pulse triggers the horizontal oscillator which operates as a blocking oscillator. The output waveform from the horizontal oscillator is processed through a sync shaper (Q12) and a clipper diode (CR5) before the resulting horizontal sync pulse is applied to the modulator diode (CR6).

Next, we observe in Fig. 11–8 that the 189-kHz output from transistor Q2 is also applied to another waveshaper which operates with transistor Q8. This shaper produces 12 rectangular pulses per horizontal-scanning interval. These pulses are applied to the color-gate diode and serve to switch the diode alternately into conduction and past cutoff. Notice that the output from the crystal-controlled chroma oscillator Q13 is also fed to the color-gate diode. Accordingly, the output from the gate diode consists of 12 color bursts per horizontal-scanning interval. This burst train is fed with the output waveform from the sync shaper to the clipper diode CR5, and thence to modulator diode CR6. The comparatively high amplitude of the horizontal sync pulse blanks the color-burst signal that occurs simultaneously, so that the waveform applied to the modulator diode CR6 is the waveform shown in Fig. 11–6.

It is not essential to provide a vertical-sync pulse in a keyed-rainbow signal. That is, if a vertical sync pulse is omitted, the chroma bars will be displayed in the usual manner, except that the bars will be observed to "roll" vertically unless the vertical-hold control in the receiver is critically

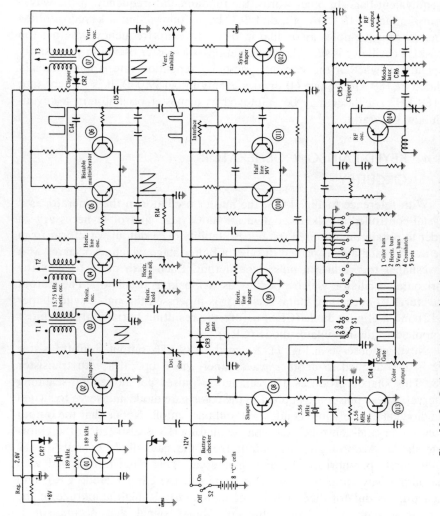

Fig. 11–8 Basic circuitry employed in a transistorized keyed-rainbow generator

adjusted. In theory, it is inconsequential whether the chroma bars roll or whether they are locked vertically. However, in practice, it is desirable to prevent vertical rolling, since the bars usually contain sufficient 60-Hz and 120-Hz residual ripple from the power supply that they produce a slight "snaking" motion if they are permitted to roll.

The vertical blocking oscillator (Q7) in Fig. 11–8 operates at the end of a frequency-divider chain that counts down 3150 times from 189-kHz to 60 Hz. Thereby, the vertical oscillator is harmonically locked to the master-oscillator frequency. This locked operation establishes an accurate relation between the horizontal and vertical sync repetition rates of 15,750 Hz and 60 Hz, so that alternate fields terminate on half-lines, and good interlacing is obtained. The vertical sync pulse generated by Q7 is simply a wide rectangular pulse; it contains no serrations and no equalizing pulses. That is, a simulated vertical sync pulse is developed in this type of generator.

Since a simulated vertical sync pulse is utilized, the horizontal oscillator in the color-TV receiver necessarily "free-runs" for the duration of the vertical sync pulse. From a practical viewpoint, this means that the horizontal oscillator may drift slightly off-frequency while the vertical sync pulse is passing through the receiver circuits. In turn, the top inch or two of a keyed-rainbow pattern may appear to be bent to the left or right, when this type of generator is used. The operator can straighten the top of the rainbow pattern by careful adjustment of the horizontal-hold control in the receiver.

As in the case of the horizontal sync pulse, the comparatively high amplitude of the vertical sync pulse blanks the color bursts that are applied simultaneously with the pulse to modulator diode CR6 in Fig. 11–8. Note that the dot-gate section and the line-gate section of the generator are both grounded during operation of the color-bar section. That is, a keyed-rainbow pattern is provided with the function switch shown set to the position in the diagram, but the dot and line signals are eliminated in this switch position. No attenuator is provided, and the RF output level is established at approximately 50,000 microvolts.

11.5   NTSC COLOR-BAR GENERATOR CIRCUITRY

It has been noted previously that NTSC color-bar generators provide various color sequences and a variety of color-difference signals. For example, the color-bar pattern shown in Fig. 11–2 is commonly provided by NTSC generators used in the service trade. This type of generator will also provide various color-difference patterns, as shown in Fig. 11–3.

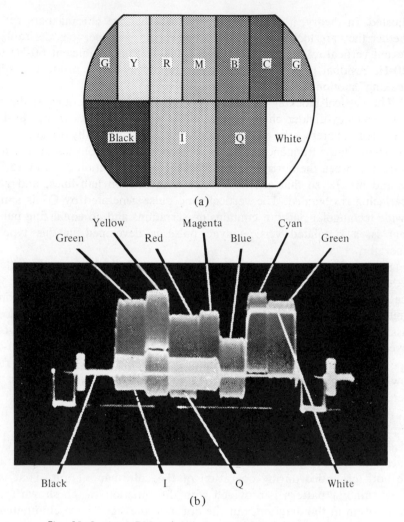

Fig. 11–9 (a) NTSC color test pattern used by transmitter engineers; (b) signal waveform

Transmitter engineers often employ a test pattern that provides the primary colors, complementary colors, I and Q color-difference signals, and the white and black levels simultaneously, as shown in Fig. 11–9(a). The signal that corresponds to this color test pattern is shown in Fig. 11–9(b).

The circuitry of an NTSC-type color-bar generator is somewhat more elaborate than that of a keyed-rainbow generator. The chroma components of an NTSC signal have different amplitudes as well as different phases; these relations are seen in Figs. 11–10 and 11–11. We will find that hue is determined by chroma phase, and saturation determined by chroma amplitude. Brightness is determined by the amplitude of the Y signal, shown in Fig. 11–10. The block diagram in Fig. 11–12 indicates the basic circuit sections employed in a typical NTSC generator. The master oscillator, or timer oscillator, operates at 315 kHz. Thus, it supplies 20 trigger pulses during each horizontal scanning interval. As explained

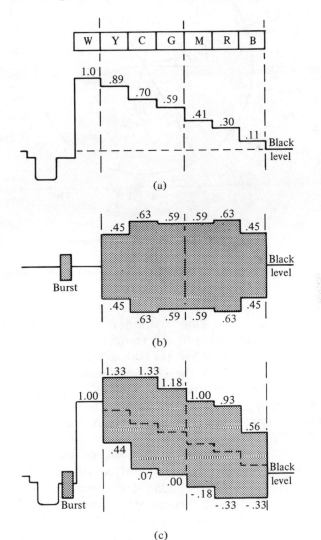

Fig. 11–10   NTSC signal components: (a) luminance or Y signal; (b) chroma signal; (c) complete color-bar signal

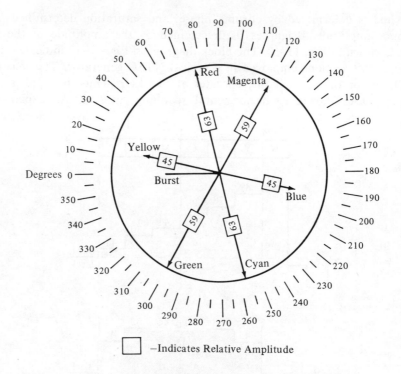

-Indicates Relative Amplitude

Fig. 11—11   Phases of the primary and complementary colors

previously, frequency-divider circuits are utilized to lock lower-frequency oscillators to the master-oscillator frequency. This counter action extends down to 60 Hz for generation of the vertical sync pulse.

We know that each color bar has both a Y component and a chroma component. With respect to the color-bar sequence illustrated in Fig. 11–2, the relative Y levels are shown in Fig. 11–13. Note that the Y levels are formed by combinations of three rectangular waveforms having comparative levels of 0.59, 0.30, and 0.11. A horizontal sync pulse is added to the Y signal in the sync-adder section (Fig. 11–12). A vertical-sync pulse is not utilized in the color-bar signal produced by this particular generator. In turn, the color bars will "roll" vertically on the picture-tube screen, unless the horizontal-hold control is carefully adjusted.

Next, the color burst and chroma signals are obtained from the 3.58-MHz crystal-controlled oscillator indicated in Fig. 11–12. Buffer action is provided by a cathode follower, and the output signal is applied to the burst-gate section and to a color-bar delay line. This delay line introduces phase shifts in the 3.58-MHz signal corresponding to the phases of the primary and complementary colors. The configuration of a chroma delay line is shown in Fig. 11–14. At the input end of the line, the signal has the B-Y phase, established by the fact that this signal is 180° out of phase with the signal output from the burst-gate section. As the signal flows down the delay line, the resulting delays provide tap-off points for the blue, cyan, green, burst, yellow, I, red, R-Y, magenta, and Q phases. This phase sequence is evident from the vector diagram in Fig. 11–15.

Additional taps may be provided on the chroma delay line if a generator supplies G-Y and G-Y$\underline{/90°}$ signals. The G-Y phase is almost the same as the green phase; G-Y$\underline{/90°}$ has a phase between yellow and I. We call G-Y and G-Y/90° quadrature signals, because they are 90° apart in phase. Similarly, R-Y and B-Y are quadrature signals, I and Q are quadrature signals. When the chroma component is added to the Y component in the composite chroma and Y adder (Fig. 11–12), the saturated color-bar signal is formed, as shown in Fig. 11–16. Again, if the generator is switched to its color-difference bar outputs, the signals have the waveforms shown in Fig. 11–17. Note that color-difference signals are centered

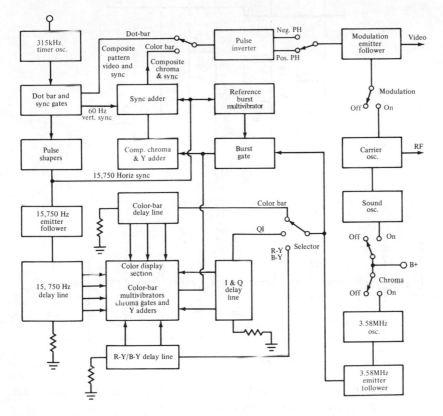

Fig. 11–12   Block diagram for a typical NTSC generator

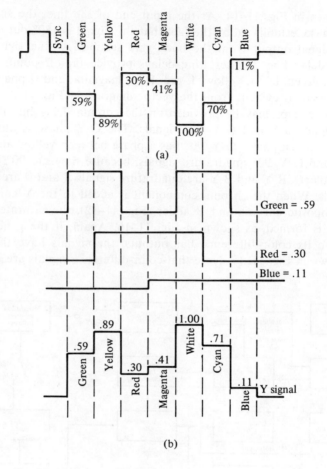

(a)

(b)

Fig. 11–13 Y component of NTSC signal

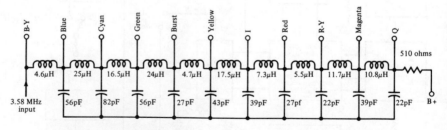

Fig. 11–14 Chroma delay line

on the black level and have no Y component. Thus, the color-difference signals are comparable with keyed-rainbow signals.

The video-frequency signals that have been described are available from the output of the cathode follower indicated in Fig. 11–12. A polarity

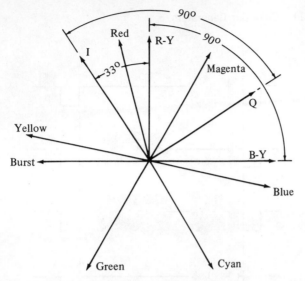

Fig. 11–15   R-Y, B-Y, I, and Q phases

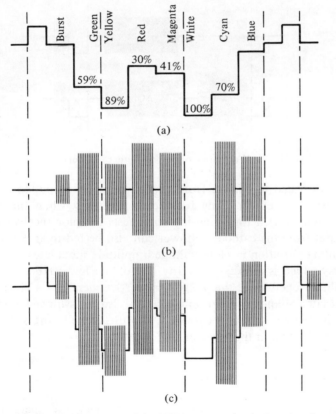

Fig. 11–16   Formation of the NTSC saturated color-bar signal:
(a) Y signal; (b) chroma signal; (c) saturated
color-bar signal

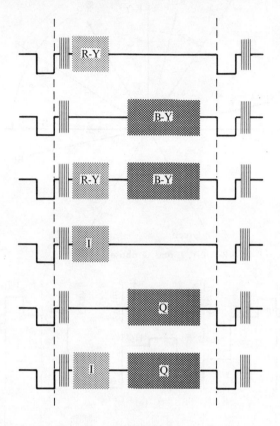

Fig. 11–17   Waveforms of color-difference signals

switch is provided to change from a positive-going output to a negative-going output waveform. This feature is necessary when the video-frequency signal is used for signal-injection tests in color receiver circuitry. The output from the cathode follower can also be fed to an RF oscillator; the resulting variation in plate voltage amplitude modulates the oscillator frequency which is adjustable to any one of the low VHF channels. The RF oscillator operates at the picture-carrier frequency of a given channel. Notice that a sound-carrier oscillator can also be switched into the RF system, if desired. A sound carrier is useful to check the setting of the fine-tuning control in the receiver under test.

11.6   BASIC APPLICATIONS

Color-bar generators are used mainly to check the operation of the chroma circuitry in color-TV receivers. Waveshapes and amplitudes

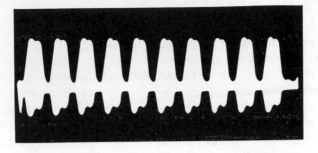

Fig. 11–18   Keyed-rainbow waveform at bandpass-amplifier
output

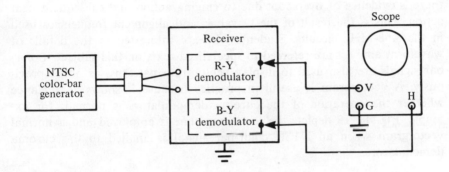

Green	Yellow	Red	Magenta	White	Cyan	Blue	Black
(8)	(1)	(2)	(3)	(4)	(5)	(6)	(7)	(8)

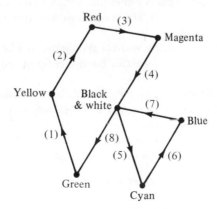

Fig. 11–19   Check of chroma-demodulator action by means
of a vectorgram display

provide information concerning circuit operation, and waveform analysis enables the experienced technician to localize defective components. Figure 11–18 illustrates the keyed-rainbow output waveform from a band-pass amplifier. We observe that nonlinear amplification is occurring because the positive peaks have greater amplitude than the negative peaks in the waveform. Although the rise and fall of the leading and trailing edges are considerably slower than in the waveform supplied by the generator, the rise and fall times are within normal tolerance with respect to standard bandpass-amplifier bandwidth.

We observe in Fig. 11–18 that the tops of the bar signals are not flat; there is evidence of overshoot due to ringing action and subsequent beat action. This is the result of the Q values and alignment frequencies used in the 3.58-MHz circuits. Students who are interested in the details of waveform analysis are referred to specialized texts on this subject. Color-bar signals are also used to display vectorgram patterns, as noted previously. A vectorgram is useful in practice because it shows at a glance whether the operation of the chroma demodulators is normal; for example, Fig. 11–19 depicts the test setup that is employed and a normal vectorgram when an NTSC color-bar signal is applied to the chroma demodulators.

QUESTIONS AND PROBLEMS

True-False

1. Two basic types of color-bar generators are in general use.
2. The NTSC generator provides true primary colors, but not true secondary colors.
3. The unkeyed-rainbow signal is sometimes called a color-display signal.
4. The first and last color bars are sometimes lost due to the long flyback time.
5. It is not necessary to provide a vertical-sync pulse in a keyed-rainbow signal.
6. The top of the rainbow pattern can be straightened out by careful adjustment of the vertical hold control.
7. Color-difference signals are comparable with keyed-rainbow signals.
8. Color-bar generators are used mainly to check the operation of the chroma circuitry in color-TV receivers.
9. In Fig. 11–18, nonlinear amplification is occurring in the amplifiers.
10. A vectorgram is most useful to test convergence.

Multiple-Choice

1. The color-bar generator most often used in service trade is the
   (a) NTSC-type of generator.
   (b) keyed-rainbow.
   (c) flying-spot scanner.

2. The output from a keyed-rainbow generator generates a color spectrum on the screen called the
   (a) bar pattern.
   (b) rainbow pattern.
   (c) ring pattern.
3. The frequency of the color burst is approximately
   (a) 60 Hz.
   (b) 15,750 Hz.
   (c) 3.6 MHz.
4. Part of the first and tenth bar may be invisible on the picture screen, due to
   (a) low frequency loss.
   (b) high frequency loss.
   (c) flyback time.
5. In practice, it is desirable to prevent vertical rolling, because of the
   (a) loss of contrast.
   (b) snaking action caused by the 60 Hz hum voltage.
   (c) color fading.
6. Color-bar generators are used mainly to check the operation of the _____ in color-TV receivers.
   (a) RF
   (b) video
   (c) chroma circuitry
7. In Fig. 11–18, the tops of the bar signals are not flat; this means that
   (a) there is a ringing action.
   (b) the circuit has a low Q.
   (c) the horizontal oscillator is free-running.
8. A vectorgram is useful in practice, because it shows at a glance whether the operation of the _____ is normal.
   (a) RF amplifier
   (b) video amplifier
   (c) chroma demodulator

General

1. What are the two basic types of color-bar generators in general use, and where are they most often utilized?
2. Which of the two types of color-bar generators provides a linear phase sweep of the chroma subcarrier?
3. What does the abbreviation NTSC mean, and what is the function of the body?
4. What type of display does an NTSC generator provide on a color television screen?
5. In respect to Fig. 11–4(a), when is full output obtained from the R-Y demodulator? When is zero output obtained from the R-Y demodulator?
6. In respect to Fig. 11–4(a), when is full output obtained from the B-Y demodulator? When is zero output obtained from the B-Y demodulator?
7. What is the frequency of the color burst signal?
8. Explain the development of the linear phase sweep signal shown in Fig. 11–5.

9. The keyed-rainbow signal comprises 11 bursts of signal. Why are only 10 chroma bars displayed in the pattern?

10. Why is it not practical to design a keyed-rainbow generator to display more than 10 color bars?

11. What is the difference between the B-Y and the R-Y signals in Fig. 11–17?

12. What is the operating frequency of the master oscillator in the keyed-rainbow generator circuit shown in Fig. 11–8?

13. What is the purpose of the buffer stage Q2 in the diagram shown in Fig. 11–8?

14. How often is the horizontal oscillator triggered in the keyed-rainbow generator, in respect to the master oscillator output?

15. Why are vertical-sync pulses developed in the keyed-rainbow generator shown in Fig. 11–8?

16. The top inch or two of a keyed-rainbow pattern may appear to be bent when a keyed-rainbow generator is used to align a color-TV. What causes this distortion and how can it be corrected?

17. What is the approximate level of the RF signal from the generator shown in Fig. 11–8?

18. What factors determine the hue and the saturation of a chroma signal?

19. What factor determines the brightness of a color display?

20. When the NTSC generator shown in Fig. 11–12 is used to align a receiver, the display on the screen will roll vertically. Why?

21. What is the purpose of the delay line in the NTSC generator shown in Fig. 11–12?

22. Why are the color-difference signals shown in Fig. 11–17 comparable with the keyed-rainbow signals?

23. What is the purpose of the polarity switch in the color generator shown in Fig. 11–12?

24. What is the purpose of the sound-carrier oscillator in the generator circuit shown in Fig. 11–12?

25. The keyed-rainbow waveform shown in Fig. 11–18 is the output from a bandpass amplifier. What characteristics of the display indicate nonlinearity of the amplifier?

26. Why are the tops of the signals shown in Fig. 11–18 rounded?

27. Draw a block diagram for the connection of a receiver and generator to check chroma-demodulator action by means of a vectorgram display.

12

Sine-Wave
and Square-Wave
Generators

12.1 SINE- AND SQUARE-WAVE
GENERATOR REQUIREMENTS

Sine-wave generators used in servicing procedures must cover the audio-frequency range. For example, high-fidelity amplifiers generally have a rated frequency response from 20 Hz to 20 kHz. Most sine-wave generators provide extended high-frequency response into the ultrasonic region, and some generators supply coverage up to 1 or 2 MHz. Service-type sine-wave generators have a maximum output level of several volts. As exemplified in Fig. 12–1, an output level of 2 volts is adequate to drive a typical audio system. It is helpful, although not essential, to have a constant output level over the audio-frequency range. Low distortion is absolutely necessary, if the sine-wave generator is to be applied in distortion measurements of high-fidelity systems. Although no industry standards have been established, it is generally agreed that high fidelity response entails a harmonic-distortion value of less than 1%. If a sine-wave has no more than 0.25% distortion over the audio-frequency range, it is satisfactory for distortion measurements.

Square-wave generators used in servicing procedures should have a repetition rate from at least 20 Hz to 100 kHz. This range is adequate for making square-wave response tests of video-frequency amplifiers in television receivers. Extended low- and high-frequency output can be useful in some applications, such as checking the response of DC and wide-band oscilloscopes. Note that the rise time of the square-wave output should not be greater than 0.08 μsec if the generator is to be used for checking the transient response of video-frequency amplifiers in TV receivers. Rise time is defined as the elapsed time between the 10% and

90% of maximum amplitude points on the leading edge of a square wave, as shown in Fig. 12–2. Rise time is measured with a triggered-sweep scope that has calibrated time-base controls. Note also that the square-wave output should have negligible overshoot, ringing, or tilt, inasmuch as these defects are among the test data taken into account when evaluating the square-wave response of a unit or system. Figure 12–3 shows basic square-wave distortions and their interpretation.

Unit	Input min.	Input max.	Output min.	Output max.
Tuner			0(variable)	2V RMS
Phono cartridge				0.1V RMS
Tape unit			0.1V (variable)	0.5V
Preamp: Tuner input	0.1V	0.5V	0.5V	2.5V
Phono input	2.0mV	70mV	0.5V	5.0V
Tape input	0.1V	0.25V	0.5V	2.5V
Power amplifier	0.25V	0.5V		50W

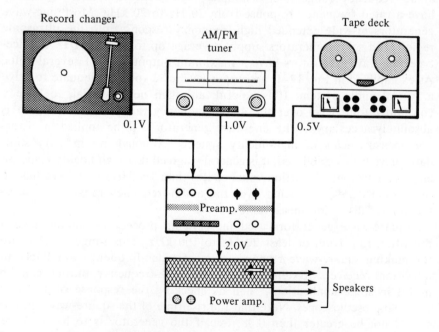

Fig. 12–1   High-fidelity audio system with operating voltage levels

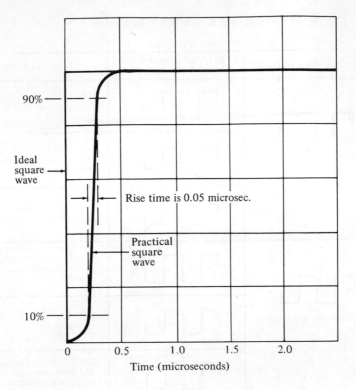

90% —

Ideal
square ——
wave

|←—→|←— Rise time is 0.05 microsec.

Practical
— square
wave

10% —

0   0.5   1.0   1.5   2.0

Time (microseconds)

(a)

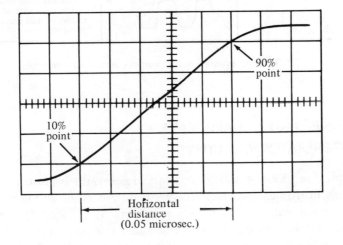

90%
point

10%
point

|←———— Horizontal ————→|
distance
(0.05 microsec.)

(b)

Fig. 12–2   Rise-time measurement on the leading edge of a
square wave: (a) pattern aspect at moderate
sweep speed; (b) pattern aspect at higher sweep
speed

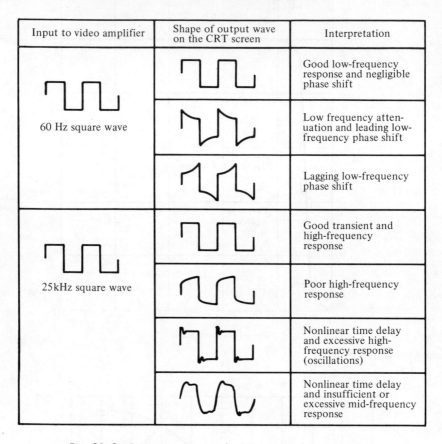

Input to video amplifier	Shape of output wave on the CRT screen	Interpretation
60 Hz square wave		Good low-frequency response and negligible phase shift
		Low frequency attenuation and leading low-frequency phase shift
		Lagging low-frequency phase shift
25kHz square wave		Good transient and high-frequency response
		Poor high-frequency response
		Nonlinear time delay and excessive high-frequency response (oscillations)
		Nonlinear time delay and insufficient or excessive mid-frequency response

Fig. 12–3   Basic square-wave distortions and their interpretation

12.2   SINE- AND SQUARE-WAVE GENERATOR CIRCUITRY

A typical service-type sine-/square-wave generator is illustrated in Fig. 12–4. This instrument provides sine-wave output from 20 Hz to 2 MHz at a maximum output level of 7.5 volts rms. It is rated for harmonic distortion of less than 0.25% over the audio-frequency range. Dial calibration is rated within ±3% from 100 Hz to 1 MHz, and within ±5% at the lower and higher ends of the frequency range. The rated amplitude variation is within ±1 dB over the audio-frequency range. Square-wave output is provided at repetition rates from 20 Hz to 200 kHz at a maximum output level of 10 volts p-p. The rise time is rated at less than 0.1

μsec at a repetition rate of 20 kHz. Thus, this type of generator meets general requirements for high-fidelity and television service applications.

Figure 12–5 shows the configuration for a sine-/square-wave generator. It consists basically of a Sulzer RC oscillator for sine-wave generation, and a bistable multivibrator for square-wave generation. The multivibrator is triggered by the sine-wave oscillator.

We observe in Fig. 12–5 that the field-effect transistor Q1 operates in the source-follower mode. This configuration provides a high input impedance. The source voltage is directly coupled to emitter-follower Q2 which buffers the signal into a regenerative feedback circuit comprising Q3, Q4, and Q5. Note that the Q4 and Q5 conduct on alternate half-cycles, and provide positive feedback to the emitter circuit of Q3 via capacitors C2 and C3, and lamp I1. Potentiometer R8 establishes the feedback level and thus controls the amplitude of oscillation. We observe that the signal voltage developed across R7 through R9 will be in phase opposition to the signal voltage across R5, thereby minimizing waveform distortion by negative-feedback action.

We also observe that the frequency of oscillation is controlled by a switch-selected bridged-T network in Fig. 12–5. This network is connected between the output of the oscillator circuit (emitter circuit of Q4 and Q5) and the gate of FET Q1. The phase-shifting characteristics of this network is such that the negative feedback applied to Q1 cancels all frequencies

Fig. 12–4   Service-type sine-/square-wave generator (Courtesy of EICO)

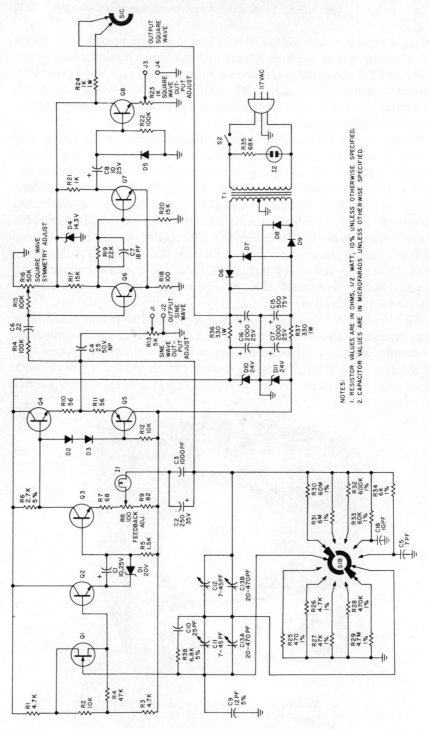

Fig. 12-5 Configuration of a sine-/square-wave generator (Courtesy of EICO)

NOTES:
1. RESISTOR VALUES ARE IN OHMS, 1/2 WATT, 10% UNLESS OTHERWISE SPECIFIED.
2. CAPACITOR VALUES ARE IN MICROFARADS UNLESS OTHERWISE SPECIFIED.

except the one that is selected by the operator. Since the bridged-T network employs 1% precision resistors, it provides accurate frequency characteristics, and a large two-gang tuning capacitor is utilized to obtain a wide frequency range. Oscillation takes place in the Q3, Q4, and Q5 sections. Lamp I1 is placed in the feedback path to stabilize the amplitude of oscillation. That is, increasing current flow through I1 causes an increase in filament resistance, which opposes the increase in current. On the other hand, less current flow is countered by reduced filament resistance. The sine-wave signal from the emitter circuit of Q4 and Q5 are coupled via C4 to potentiometer R3 which controls the output signal level.

Next, we observe that square-wave output is produced by the Schmitt trigger circuit Q6-Q7 in Fig. 12–5. Triggering is provided by sine-wave input, and conduction occurs on the positive rise of the sine wave. Thus, conduction of Q6-Q7 takes place on alternate half-cycles of the trigger voltage, and a square-wave voltage is obtained at the collector of Q7. Symmetry of the square-wave output is controlled by potentiometer R16, which establishes the DC level at the base of Q6. Zener diode CR4 is employed to regulate the collector supply voltage. From Q7, the square-wave output is coupled via C8 to emitter-follower Q8, which supplies its output to the level-control R23.

12.3 SINE-WAVE GENERATOR APPLICATIONS

Signal-substitution tests are made with a sine-wave generator to localize defects in audio systems, as depicted in Fig. 12–6. The test signal is injected at suitable levels progressively through the system until the defective section or stage is identified. A stereo system usually develops a fault in one channel only. In turn, the normally operating channel can be used as a reference for comparison of stage gains and operating voltages.

Audio-amplifier frequency-response checks are made with the test arrangement depicted in Fig. 12–7(a). Note that the speaker is disconnected from the amplifier under test, and a power resistor of suitable value is connected across the amplifier output terminals. This procedure eliminates the objectionable noise that would be produced otherwise during the test. The amplifier is energized by a sine-wave generator, and a VOM or TVM is utilized as an output indicator. A high-fidelity amplifier normally has less than ±1 dB amplitude variation over a frequency range from approximately 20 Hz to 20 kHz. Note in Fig. 12–7(b) that the bass and treble tone controls must be set to their midpoints, or the effective frequency response will be considerably affected. Note that equalizer

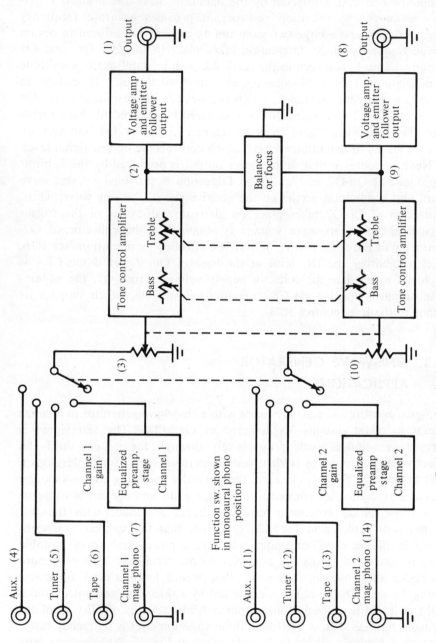

Fig. 12–6 Signal-substitution tests in a stereo system

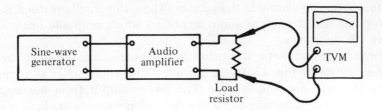

(a)

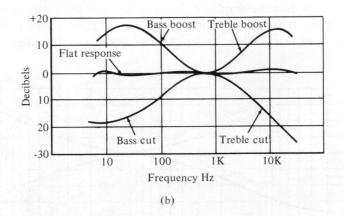

(b)

Fig. 12–7   Check of audio-amplifier frequency response: (a)
test connections; (b) effect of tone-control settings

frequency characteristics also alter the effective frequency response at various inputs of a hi-fi amplifier, as exemplified in Fig. 12–8.

A simple test for audio-amplifier distortion is shown in Fig. 12–9. This arrangement produces Lissajous patterns on the scope screen. Examples of pattern evaluation are noted in the diagram. It is advantageous to make a distortion check at maximum rated power output of the amplifier, since distortion generally increases with an increase in power output. However, crossover distortion does not increase, because this fault is greatest at low power levels. Note that although this test method relaxes the demand for low-distortion output from the sine-wave generator, it requires high-performance vertical and horizontal amplifiers in the scope. That is, although generator distortion cancels out in the test, any scope distortion will be falsely charged against the amplifier under test. Not all service scopes can meet the requirements imposed by this test method.

We will find that it is comparatively difficult to measure percentage harmonic distortion with the test method depicted in Fig. 12–9. Even if the scope employs high-fidelity vertical and horizontal amplifiers, the amount of pattern change caused by 1% distortion, for example, is difficult to observe. However, if the distortion amounts to several per cent, the fact shows up clearly in the pattern. Thus, this simple method is most appropriate for utility-type audio amplifiers which normally operate well above 1% distortion.

Percentage harmonic distortion is accurately measured in the test method shown in Fig. 12–10. A harmonic-distortion meter is connected across the amplifier load resistor. The power output from the amplifier can be determined by measuring the AC voltage across the load resistor. Note that the sine-wave generator used in this test must have substantially less distortion than the amplifier. The first test is made at maximum rated power output from the amplifier, inasmuch as most forms of distortion show up most prominently under this condition. On the other hand, the

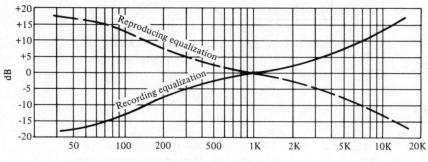

(a)

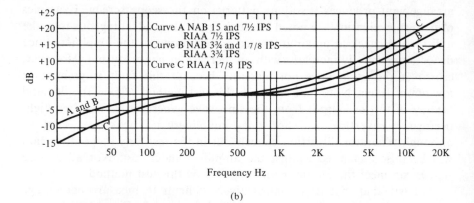

(b)

Fig. 12–8   Equalizer frequency characteristics: (a) RIAA standard reproducing and recording equalization curves for discs; (b) standard playback equalization curves for tapes

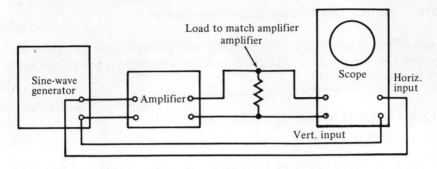

(a)

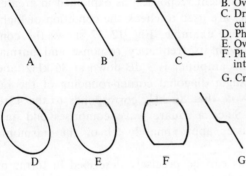
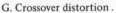

A. No overload distortion, no phase shift.
B. Overload distortion, no phase shift.
C. Driving into base current and past cut-off, no phase shift.
D. Phase shift
E. Overload distortion and phase shift
F. Phase shift, driving into cut-off and into saturation.
G. Crossover distortion.

(b)

Fig. 12–9   Simple distortion test arrangement: (a) test connections; (b) pattern evaluation

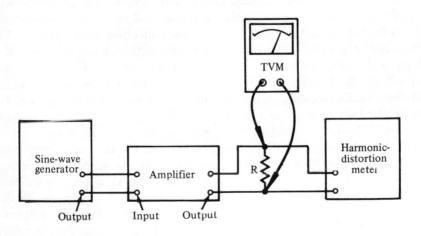

Fig. 12–10   Measurement of percentage harmonic distortion

second test is made at a comparatively low power output from the amplifier, because crossover distortion develops a maximum percentage value in low-level operation.

12.4   SQUARE-WAVE GENERATOR APPLICATIONS

A square wave can be built up from a very large number of sine waves as depicted in Fig. 12–11. This viewpoint is helpful in evaluating some of the results obtained in square-wave tests. However, we should recognize that a more useful evaluation can usually be obtained from the viewpoint of exponential waveforms and transient responses, as explained in greater detail subsequently. Square waves are used to check the condition of high-fidelity amplifiers and systems. For example, Fig. 12–12 shows the configuration for a hi-fi amplifier with its frequency response and normal 2-kHz square-wave response. This amplifier is 3 dB down at 86 kHz, and we observe that it produces a slight diagonal corner-rounding of the reproduced 2-kHz square wave. Note that 86 kHz corresponds to the 43rd harmonic of the square wave. Since a square wave comprises odd harmonics only, the amplifier passes approximately 21 of the harmonic components in this example.

Many square-wave distortions can be precisely described in terms of exponential waveforms. The basic exponential waveform is called the natural law of growth and decay. A simple example is depicted in Fig. 12–13(a). When a capacitor is charged and discharged through a resistor, the exponential curves shown in Fig. 12–13(b) are displayed on the scope screen. These waveforms are presented in a chart framework that is called the universal time-current chart. This means that all series RC circuits have the same exponential response, and that the time-constant of a series RC circuit is equal to the product of ohms times farads, with the answer expressed in seconds. That is, the capacitor will attain 63.2% of its final charge in one time-constant, or RC seconds. Conversely, the capacitor will discharge to 38.8% of its initial voltage in one time-constant, or RC seconds. After five time-constants, the capacitor will be practically fully charged, or fully discharged.

In Fig. 12–13, it is evident that since the charge or discharge current flows through resistor R, that curve B represents the resistor voltage drop during charge or discharge of the capacitor. If the scope is connected across the capacitor, we observe waveform A during charge and waveform B during discharge. This mode of operation is called integration. If the scope is connected across the resistor, on the other hand, we observe waveform B during charge or discharge. This mode of operation is called

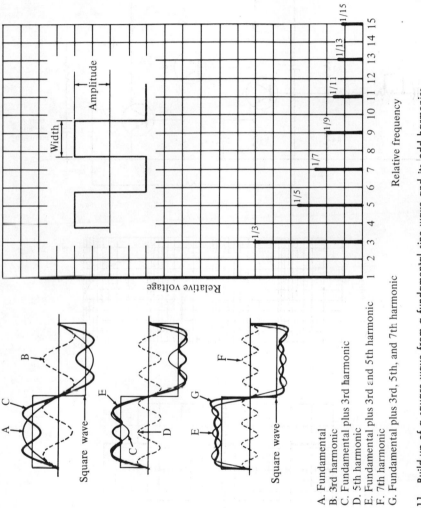

A. Fundamental
B. 3rd harmonic
C. Fundamental plus 3rd harmonic
D. 5th harmonic
E. Fundamental plus 3rd and 5th harmonic
F. 7th harmonic
G. Fundamental plus 3rd, 5th, and 7th harmonic

Fig. 12–11 Build-up of a square wave from a fundamental sine wave and its odd harmonics

271

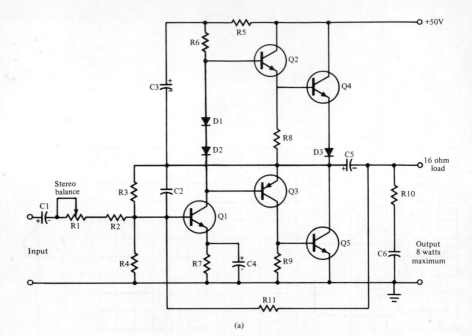

(a)

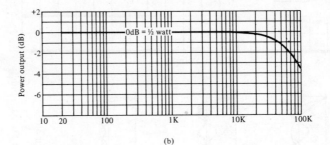

(b)

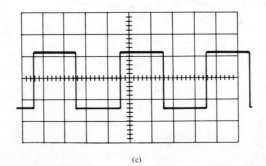

(c)

Fig. 12—12   Square-wave response of a high-fidelity amplifier:
(a) amplifier configuration; (b) normal frequency
response; (c) normal 2-kHz square-wave response
(Courtesy of General Electric)

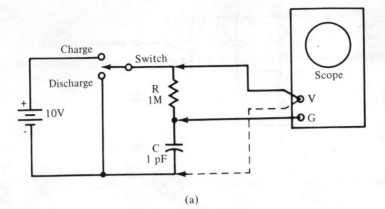

(a)

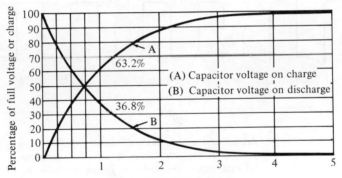

(b)

Fig. 12–13 Basic exponential waveforms: (a) series RC charging and discharging circuit; (b) exponential waveforms in universal time-constant chart form

differentiation. Figure 12–14 shows a practical application of these principles. In (a), we see a simple compensated step attenuator such as is used
in an oscilloscope. If trimmer capacitor C1 is correctly adjusted, an un-
distorted square wave is reproduced. On the other hand, if C1 is misadjusted, we will observe the 10-kHz square-wave distortions shown in (b).

It is instructive to observe how the waveforms are related around a
series RC circuit. Figure 12–15 shows that the waveform across the resistor adds to the waveform across the capacitor to form the applied
square wave. This result, of course, is required by Kirchhoff's law. In
practical circuitry, we are often concerned with successive exponential
operations on an applied square wave. For example, Fig. 12–16 shows a

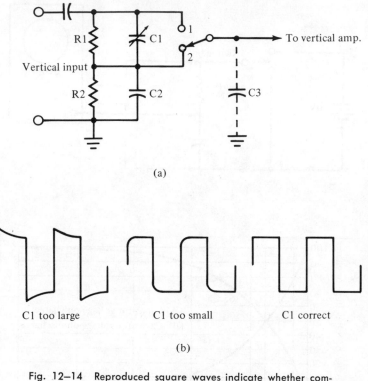

(a)

C1 too large   C1 too small   C1 correct

(b)

Fig. 12–14   Reproduced square waves indicate whether compensated step attenuator is adjusted properly:
(a) attenuator configuration; (b) square-wave reproduction with three settings of C1

universal time-constant chart for one-, two-, and three-section integrators.
These are symmetrical networks in which each integrating section is the
same. Two- and three-section integrators are used in various TV receiver
circuits. Universal time-constant charts are useful because the calculations
that are required to find the normal transient response of various RC
circuits are quite difficult. On the other hand, the answer is given directly
in the chart presentation.

    We will find that the bandwidth of an amplifier is directly related to
the rise time of a square wave reproduced at the output of the amplifier.
Bandwidth is defined as the frequency at which the amplifier response is 3
dB down. This is called the cut-off frequency of the amplifier. The cut-off
frequency and rise time are related as follows:

$$f_{co} = \frac{1}{3T}$$

where T is the rise time of the reproduced waveform.

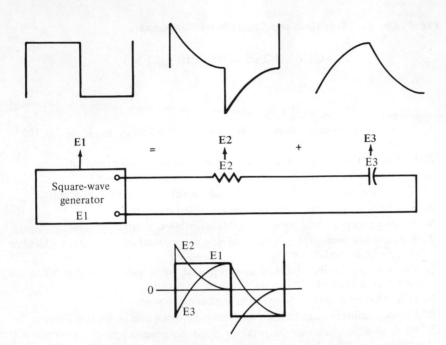

Fig. 12–15 Kirchhoff's law applied to the square-wave response of a series RC circuit

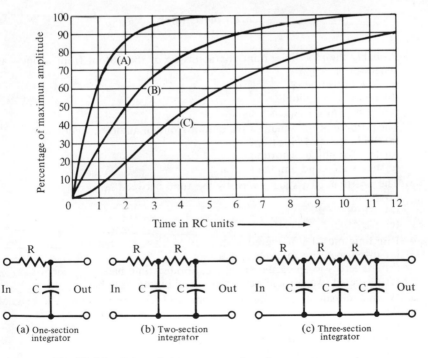

Fig. 12–16 Universal time-constant chart for one-, two-, and three-section symmetrical integrator circuits

QUESTIONS AND PROBLEMS

True-False

1. Most sine-wave generators are limited to a frequency range of 20 Hz to 20 kHz.
2. Low distortion of the waveform is essential in an audio generator.
3. The square-wave output can be used to test video amplifiers.
4. Rise time is defined as the time elapsed between 0 and 90% of a pulse.
5. Rise time is measured with a triggered-sweep oscilloscope.
6. A square wave should have negligible overshoot, ringing, or tilt.
7. A sine-wave generator can be used for a signal-substitution test to localize defects in an audio system.
8. The speaker is disconnected and replaced by a power resistor when an amplifier is under test for a more linear response.
9. Distortion tests should be made at maximum power.
10. Lissajous patterns can be used to determine distortion of an amplifier.
11. Most oscilloscopes can be used to measure percentage of harmonic distortion.
12. A square wave is made up of many sine waves.

Multiple-Choice

1. Probably the most important characteristic of the output from a sine-wave generator is
   (a) low distortion.
   (b) constant amplitude.
   (c) an accurate frequency dial.
2. The purpose of the square-wave output from an audio oscillator is to
   (a) calibrate an oscilloscope.
   (b) test for phase distortion.
   (c) test wide-band amplifiers.
3. The rise time of a square wave is the time between ____% and ____%.
   (a) 0; 90
   (b) 10; 90
   (c) 50; 50
4. Rise time must be measured with a _____ that has _____
   (a) dwell meter; a microsecond scale.
   (b) triggered-sweep oscilloscope; a calibrated time base.
   (c) triggered-sweep oscilloscope; an uncalibrated time base.
5. When audio-amplifier frequency-response checks are made with test signals, the
   (a) speaker is disconnected.
   (b) speaker is replaced by a power resistor.
   (c) speaker must be connected.

6. Maximum distortion usually is at the
   (a) maximum power level.
   (b) minimum power level.
   (c) mid-frequency range.
7. Percentage harmonic distortion is accurately measured with
   (a) a TVM.
   (b) an oscilloscope.
   (c) a harmonic-distortion meter.
8. A square wave can be used for
   (a) high-frequency response tests.
   (b) low-frequency response tests.
   (c) both (a) and (b).
9. If an oscilloscope is connected across a capacitor to measure the RC time, the test is called
   (a) a capacitor test.
   (b) a differentiator test.
   (c) an integrator test.
10. The cutoff frequency and rise time of a circuit are related by the formula
    (a) $f_{co} = 3T$.
    (b) $f_{co} = 1/3T$.
    (c) $f_{co} = 13/T$.

General

1. What is the frequency range of sine-wave generators?
2. What are the requirements of a quality square-wave generator?
3. What are some of the defects that can be located in an audio system by a sine-wave generator?
4. When testing an audio amplifier with a signal generator, why is the speaker replaced with a load resistor?
5. How is percentage of harmonic distortion usually measured?
6. How is a square wave used to evaluate the operation of an audio amplifier?
7. Suppose the rise time of a square wave at the output of an amplifier is 20 μsec. What is the approximate frequency response of the amplifier?

Ind